T0140023

Submarine Hydrodynamics

Martin Renilson

Submarine Hydrodynamics

Second Edition

Martin Renilson
Australian Maritime College
University of Tasmania
Launceston, Tasmania
Australia

ISBN 978-3-030-07716-7 ISBN 978-3-319-79057-2 (eBook)
https://doi.org/10.1007/978-3-319-79057-2

Softcover re-print of the Hardcover 2nd edition 2018

Printed on acid-free paper

This Springer imprint is published by the registered company Springer International Publishing AG part of Springer Nature
The registered company address is: Gewerbestrasse 11, 6330 Cham, Switzerland

This book is dedicated to my wonderful wife, Susan, who has looked after me when I have been ill. She has also been a great support and has assisted with the arrangement of the manuscript. I would not have managed to have completed this without her.

Acknowledgements

A number of people have assisted me greatly with the preparation of this book, and it is not possible to mention all of them by name. However, I am particularly indebted to Brendon Anderson; Paul Blythe; Matteo Bonci; Bruce Cartwright; Paul Crossland; Ian Dand; Giulio Dubbioso; Jon Duffy; Eric Fusil; Zhi Leong; Mohammad Moonesun; Chris Polis; David Pook; Dev Ranmuthugala; Amit Ray; Chris Richardsen; Prasanta Sahoo; Greg Seil; Debabrata Sen; Auke van der Ploeg; Serge Toxopeus; and George Watt for providing information that I have made use of.

I am also very grateful to my wife, Susan Renilson, for all the effort that she put in correcting my English, pointing out where my explanations make no sense, and checking for consistency throughout this whole book.

Martin Renilson

Contents

About the Author

Martin Renilson (Prof.) has been working in the field of Ship Hydrodynamics for over 35 years. He established the Ship Hydrodynamics Centre at the Australian Maritime College (AMC) in 1983 and was Director of the Australian Maritime Engineering Cooperative Research Centre in 1992. He started the Department of Naval Architecture & Ocean Engineering at AMC in 1996, which he ran until 2001 when he was appointed Technical Manager, Maritime Platforms & Equipment for DERA/QinetiQ in the UK.

In 2007, he returned to Australia and set up his own company, conducting maritime-related consulting. He also held a part-time chair in hydrodynamics at AMC, now an institute of the University of Tasmania.

In 2012, he was appointed inaugural Dean of Maritime Programs at the Higher Colleges of Technology, United Arab Emirates, to start maritime education for the country. He retired from this position in November 2015 and returned to Tasmania.

He holds the position of Adjunct Professor in Hydrodynamics at the University of Tasmania, Australia and is President of the Australian Division of the Royal Institution of Naval Architects.

E-mail: martin@renilson-marine.com

Nomenclature and Abbreviations

Notes

1. Where possible, the notation used for manoeuvring is the same as that given by Gertler and Hagen (1967), however much of that is repeated here for completeness.
2. The body-fixed axis system is given in the figure below. The origin, O, is taken on the centreline at the position of the longitudinal centre of gravity of the submarine. The positive linear distances, velocities, accelerations and forces are all in the positive direction of the relevant axes, and the positive rotational values are all in the clockwise direction looking along the positive direction of the axes from the origin.
3. The prime notation is used for non-dimensionalisation, where non-dimensional quantities are denoted by a dash, as with: X', indicating the non-dimensional form of the force in the longitudinal axis, X. Unless otherwise stated, non-dimensionalisation is achieved by dividing the quantity by ½ density of water times length and velocity to the required powers.
4. Coefficients of forces and moments when manoeuvring are denoted by subscripts referring to the velocities and accelerations which the relevant force, or moment, is a function of. For example, Y_v denotes the first-order coefficient used in representing the sway force, Y, as a function of sway velocity, v. This is the partial derivative of the sway force, Y, with respect to sway velocity, v.
5. Differentiation with respect to time is denoted by a dot above the variable. For example, $\dot{v}$ is the derivative of sway velocity with respect to time—the sway acceleration.
6. Nonlinear coefficients of forces and moments, and those due to coupling, are represented by the relevant subscripts. For example, the nonlinear coefficient of sway force, Y, as a function of sway velocity, v, is represented by: $Y_{v|v|}$. Note that in this case the modulus of the sway velocity is used because the function is an odd function.

7. Where possible, the notation used is that commonly used for the topic being discussed. Thus, in some cases, the same quantity is defined by different symbols in different chapters.
8. For brevity, where symbols are only used in one location in the text and are clearly defined there, then these are not always defined in the notation.

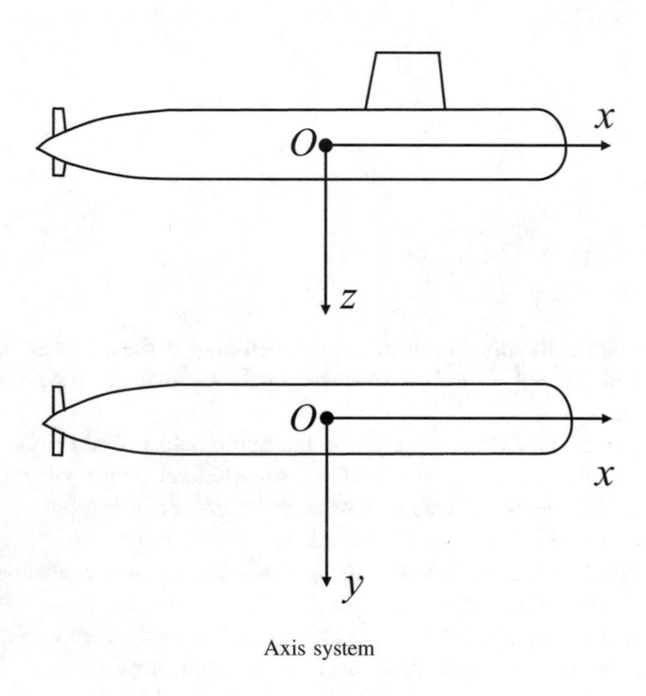

Axis system

Symbols

$A_{frontal}$	Sail frontal area
A_F	Fore body frontal area
A_m	Submarine midships cross-sectional area
A_{plan}	Plan area of appendage
A_{wind}	Profile windage area above the waterline
a	Chord of flat plate
a_i, b_i, c_i	Coefficients used to represent the resistance of the submarine in the x-axis
B	Upward force due to the buoyancy = $\nabla\rho g$
B	Position of centre of buoyancy
B_F	Position of centre of buoyancy of form displacement

BG	Distance between the centre of buoyancy and the centre of gravity
BG_F	Distance between the centre of buoyancy and the centre of gravity corrected for free surface
B_H	Position of centre of buoyancy of hydrostatic displacement
BM	Distance between the centre of buoyancy and the metacentre
B_p	Propulsor loading coefficient
b	Span of flat plate
bg	Vertical upward force through the centre of buoyancy
C_A	Correlation allowance
C_D	Non-dimensional drag coefficient at zero angle of attack
C_{D_α}	Non-dimensional slope of drag as a function of angle of attack
C_F	Non-dimensional friction resistance coefficient = $R_F/\left(\frac{1}{2}\rho SV^2\right)$
$C_{F_{flat}}$	Non-dimensional flat plate frictional resistance = $R_{F_{flat}}/\left(\frac{1}{2}\rho SV^2\right)$
$C_{F_{form}}$	Non-dimensional frictional resistance including frictional-form resistance = $R_{F_{form}}/\left(\frac{1}{2}\rho SV^2\right)$
C_{L_α}	Non-dimensional slope of lift as a function of angle of attack
$C_{L_{\delta B}}$, $C_{L_{\delta R}}$, $C_{L_{\delta S}}$	Non-dimensional slope of the lift as a function of deflection angle for the bow plane, the rudder and the stern plane, respectively
C_p	Prismatic coefficient = $\nabla / A_m L$
C_p	Non-dimensional pressure drag = $R_P/\left(\frac{1}{2}\rho SV^2\right)$
C_p	Pressure coefficient
$C_{p_{fb}}$	Non-dimensional pressure drag on fore body
C_R	Non-dimensional residual resistance coefficient
C_T	Total resistance coefficient
c_{sail}	Chord of sail
D	Hull diameter
D	Propulsor diameter
D_{local}	Local diameter at element of propulsor
D_C	Distortion coefficient
$\bar{d}$	Diameter of equivalent ellipsoid of revolution
d_T	Diameter of the trip wire
F_D	Skin friction correction force
F_r	Froude number = $V/\sqrt{gL}$
G	Position of centre of gravity
G_F	Position of centre of gravity corrected for free surface
G_F	Position of centre of gravity of form displacement
G_H	Position of centre of gravity of hydrostatic displacement
GM	Distance between the centre of gravity and the metacentre
G_FM	Distance between the centre of gravity corrected for the free surface and the metacentre

GZ_{Max}	Maximum value of the righting lever
G_H	Stability index in the horizontal plane
G_V	Stability index in the vertical plane
g	Acceleration due to gravity
h	Vertical distance between the windage surface centre and the driftage surface centre
H	Distance from the water surface to the centreline of the submarine
H^*	Non-dimensional distance from the water surface to the centreline of the submarine = H/D
H^*	Coefficient values for use with Sen Sensitivity Index
$H_{1/3}$	Significant wave height
I	Second moment of area of the waterplane around the longitudinal axis
I_{xx}, I_{yy}, I_{zz}	Mass moments of inertia about the *x*-axis, the *y*-axis and the *z*-axis, respectively
I_{xy}, I_{yx}, I_{zx}	Products of inertia about *xy*, *yx* and *zx*, respectively
I'_{yy} and I'_{zz}	Non-dimensional moments of inertia in pitch and yaw, respectively
J	Propeller advance coefficient
J_T	Propeller advance coefficient achieved by the thrust identity method
J_Q	Propeller advance coefficient achieved by the torque identity method
K	Position of the keel
K_0, K_1	Coefficients used to define aft body of Suboff
K, M, N	Moments about the *x*-axis, the *y*-axis and the *z*-axis, respectively
K', M', N'	Non-dimensional moments about the *x*-axis, the *y*-axis and the *z*-axis, respectively = moment/($\frac{1}{2}\rho V^2 L^3$)
K_a	Coefficient of added mass
KB	Distance between the keel and the centre of buoyancy
KB_F	Distance between the keel and the centre of buoyancy of the form displacement
K_c	Casing factor
KG_F	Distance between the keel and the centre of gravity of the form displacement
KG_F	Distance between the keel and the centre of gravity corrected for free surface
KM	Distance between the keel and the metacentre
K_P	Ratio of pressure resistance to friction resistance
K_Q	Propeller torque coefficient = $Q/\rho D^5 n^2$
K_{QM}	Propeller torque coefficient measured on self-propelled model

K_{QT}	Propeller torque coefficient obtained by the thrust identity method
K_T	Propeller thrust coefficient = $T/\rho D^4 n^2$
K_{Ty}	Non-dimensional side force from the propeller
K_{TQ}	Propeller thrust coefficient obtained by the torque identity method
$K_{Q_{(J=0)}}$	Value of the torque coefficient for $J = 0$
K_{TM}	Propeller thrust coefficient measured on self-propelled model
$K'_{\delta X_i}, M'_{\delta X_i}, N'_{\delta X_i}$	Non-dimensional coefficient of moment due to the angle of appendage X_i about the x-axis, y-axis and z-axis, respectively
$K'_*, M'_*, N'_*, Y'_*, Z'_*$	Non-dimensional roll moment, pitch moment, yaw moment, sway force and heave force, respectively, when the submarine is travelling at steady state with $p = q = r = v = w = 0$ and no appendage deflection angles
k_s	Sail efficiency factor
k_{sp}	Sail plane efficiency factor
k_{WB}	Stern plane efficiency factor
k_x, k_y, k_z	Added mass coefficients for motion in the x, y, and z directions, respectively
L	Length
L_A	Length of aft body
Lbp	Length between perpendiculars
L_F	Length of fore body
Loa	Length overall
L_{PMB}	Length of parallel middle body
l_{app}	Horizontal coordinate of the centre of pressure, or centre of added mass, of an appendage
l_w	Wind heeling lever
M	Position of the metacentre
M_{in}, M_{out}	In-phase and out-of-phase components, respectively, of the measured pitch moment during a PMM test
M_{MEAN}	Mean pitch moment in waves
M'_{MEAN}	Non-dimensional mean pitch moment in waves = $M_{MEAN}/\rho g L D \zeta_w^2$
$M_m(t), Z_m(t)$	Measured pitch moment and heave force as a function of time
M_{RAO}	First–order pitch moment response amplitude operator
M'_{RAO}	Non-dimensional first-order pitch moment response amplitude operator = $M_{RAO}/\rho g L^2 D \zeta_w$
$M_{w_{app}}, M_{q_{app}}$	Rate of change of moment about the y-axis on an appendage as a function of heave velocity and pitch velocity, respectively
m	Mass of the submarine
m_{added}	Added mass

m'	Non-dimensional mass = m/(½ρL^3)
mg	Vertical downward force through the centre of gravity
N	Propulsor rate of rotation (revolutions per minute)
N_{in}, N_{out}	In-phase and out-of-phase components, respectively, of the measured yaw moment during a PMM test
$N_{v_{app}}, N_{r_{app}}$	Rate of change of moment about the z-axis on an appendage as a function of sway velocity and yaw velocity, respectively
n	Propulsor rate of rotation (revolutions per second)
n	Power for fineness factor (Eq. 4.19)
n_{PMB}	Power for parallel middle body factor (Eq. 4.19)
n_f	Coefficient defining the fullness of the fore body
O	Position of the origin
P	External vertical force due to grounding or contact with ice
P_B	Brake power (from engine)
P_E	Effective power
P_S	Shaft power
P_T	Thrust power
p, q, r	Angular velocities about the *x*-axis, the *y*-axis and the *z*-axis, respectively
$\dot{p}, \dot{q}, \dot{r}$	Angular accelerations about the *x*-axis, the *y*-axis and the *z*-axis, respectively
p', q', r'	Non-dimensional angular velocities about the *x*-axis, the *y*-axis and the *z*-axis, respectively = angular velocity × L/V
Q	Torque on propeller
Q_M	Propeller torque in self-propulsion test
R	Radius of turning circle
R^*	Manoeuvring response parameter for use with Sen Sensitivity Index
$R_{control\,surface}$	Drag of control surface
R_e	Reynolds number = VL/v
$R_{F_{flat}}$	Friction resistance of a flat plate
$R_{F_{form}}$	Frictional resistance including frictional-form resistance
R_P	Form drag
$R_{sail_{form}}$	Form drag of sail
R_T	Total resistance
r	Radius
r_h	Coefficient used in definition of Suboff aft body
r_{x_f}	Radius of the section of the fore body at a distance x_f from the rearmost part of the fore body
S	Wetted surface area
S	Sen Sensitivity Index
S_a	Planform area of lifting surface

S_{hull}	Wetted surface of submarine hull
S_{sail}	Wetted surface of sail
T	Thrust of propulsor
T_0	Wave modal period
T_M	Propeller thrust in self-propulsion test
t	Thrust deduction fraction
t	Time
t_{sail}	Thickness of sail
U_1	Streamwise velocity at the edge of the boundary layer
U_∞	Nominal streamwise velocity
u, v, w	Velocities in the x, y and z directions, respectively
$\dot{u}$, $\dot{v}$, $\dot{w}$	Accelerations in the x, y and z directions, respectively
u', v', w'	Non-dimensional velocities in the x, y and z directions, respectively = velocity/V
u_{aB}, u_{aR}, u_{aS}	Axial velocity at the bow plane, the rudder and the stern plane, respectively
u_c	Steady-state velocity in the x-axis at the set propeller rpm when the submarine has only velocity in the x-axis and has no control surfaces deflected
V	Velocity
V_a	Velocity of advance of the propulsor
V_B, V_R, V_S	Velocity at the bow plane, the rudder and the stern plane, respectively
$V_{B_{eff}}, V_{R_{eff}}, V_{S_{eff}}$	Effective velocity at the bow plane, the rudder and the stern plane, respectively
V_{eff}	Effective velocity (general)
V_{wind}	Wind speed (in knots)
V_θ	Local tangential velocity into the propulsor blade
V^*	Local axial velocity into the propulsor blade
v_R	Sway velocity at the rudder (uncorrected for the presence of the hull)
W	Downward force due to the mass = Δg
WOF	Wake Objective Function
w	Taylor wake fraction
$\bar{w}$	Average wake fraction at a given radius
w_Q	Taylor wake fraction obtained by the torque identity method
w_T	Taylor wake fraction obtained by the thrust identity method
$\mathrm{w_B} w_s$	Heave velocity at the bow plane and stern plane, respectively (uncorrected for the presence of the hull)
X , Y, Z	Forces in the x-axis, y-axis and z-axis, respectively
$X,' Y', Z'$	Non-dimensional forces in the x-axis, y-axis and z-axis, respectively = force/($\frac{1}{2}\rho V^2 L^2$)

$X'_{\delta X\delta X_i}$, $Y'_{\delta X_i}$, $Z'_{\delta X_i}$	Non-dimensional coefficient of force due to the angle of appendage X_i in the *x*-axis, *y*-axis and *z*-axis, respectively
x, y, z	Coordinates in the *x*-axis, *y*-axis and *z*-axis, respectively
x_A	Distance in the *x* direction aft of the forward most part of the aft body
x_B, y_B, z_B	Coordinates of the centre of buoyancy in the *x*-axis, *y*-axis and *z*-axis, respectively
x_{bow}, x_{rudder}, x_{stern}	*x* coordinate of the bow plane, the rudder and the stern plane, respectively
x_{CLR}	*x* coordinate of the position of the Centre of Lateral Resistance
x_{CP}	*x* coordinate of the position of the Critical Point
x_f	Distance in the *x* direction forward of the rearmost part of the fore body
x_{SUBOFF}	Distance in the *x* direction aft of the forward perpendicular (used for the definition of the shape of the DARPA Suboff)
x_G, y_G, z_G	Coordinates of the centre of gravity in the *x*-axis, *y*-axis and *z*-axis, respectively
x'_G	Non-dimensional *x* coordinate of the position of the centre of gravity = x_G/L
x_{NP}	*x* coordinate of the position of the Neutral Point
Y_{in} Y_{out}	In-phase and out-of-phase components, respectively, of the measured sway force during a PMM test
Y_r, Y_v, Z_q, Z_w	First-order coefficients of force as functions of velocities (*q, r, v,* and *w*)
$Y_{v_{app}}$, $Y_{r_{app}}$	Rate of change of force in the *y*-axis on an appendage as a function of sway velocity and yaw velocity, respectively
$Y'_{\dot{v}_{app}}$	Contribution of an appendage to the non-dimensional sway added mass coefficient
y_0, z_0	Amplitude of oscillation in the *y*-axis and *z*-axis, respectively, during PMM tests
Z_{in}, Z_{out}	In-phase and out-of-phase components, respectively, of the measured heave force during a PMM test
Z_{MEAN}	Mean heave force in waves
Z'_{MEAN}	Non-dimensional mean heave force in waves = $Z_{MEAN}/\rho g L \zeta_w^2$
Z_{RAO}	First-order heave force response amplitude operator
Z'_{RAO}	Non-dimensional first-order heave force response amplitude operator = $Z_{RAO}/\rho g L^2 \zeta_w$
$Z_{w_{app}}$, $Z_{q_{app}}$	Rate of change of force in the *z*-axis on an appendage as a function of heave velocity and pitch velocity, respectively
$Z'_{\dot{w}_{app}}$	Contribution of an appendage to the non-dimensional heave added mass coefficient
α	Angle of attack

α_t	Half tailcone angle
β	Angle of flow into propeller blade
$\gamma_B, \gamma_R, \gamma_S$	Flow straightening effect of the presence of the submarine hull for the bow plane, the rudder and the stern plane, respectively
δ	Appendage deflection angle
δ_B, δ_R, δ_S	Deflection angle of the bow plane, the rudder and the stern plane, respectively
$\delta_{B_{eff}}, \delta_{R_{eff}}, \delta_{S_{eff}}$	Effective bow plane angle, rudder angle and stern plane angle, respectively
δ_0	Amplitude of rudder angle oscillation in zigzag manoeuvre
Δ	Displacement
ΔC_F	Roughness allowance
Δ_F	Form displacement
Δ_H	Hydrostatic displacement
ζ_w	Wave height
η	Ratio of self-propulsion velocity for set value of rpm to actual velocity
η	Propeller efficiency
η_B	Efficiency of propeller when behind the submarine
η_H	Hull efficiency: ratio of effective power to thrust power
η_O	Open water propeller efficiency
η_R	Relative rotative efficiency
η_{RQ}	Relative rotative efficiency obtained by the torque identity method
η_{RT}	Relative rotative efficiency obtained by the thrust identity method
η_Q	Propeller efficiency obtained by the torque identity method
η_T	Propeller efficiency obtained by the thrust identity method
θ	Pitch angle
θ_0 ψ_0	Amplitude of oscillation about the *y*-axis and *z*-axis, respectively, during PMM tests
θ_0	Angle of heel under the action of a steady wind
θ_1	Angle of roll to windward due to wave action
θ_2	Angle of downflooding (θ_F) or 50°, whichever is less
θ_c	Angle of the second intercept between the wind heeling lever (l_{w2}) and the righting lever
θ_F	Angle of downflooding
ν	Kinematic viscosity
ξ_{PMB}	Parallel middle body factor
ξ_{hull}	Hull form factor
ρ	Density of water
τ	Trim angle
ϕ	Roll angle/heel angle

ψ	Drift angle and heading angle
ψ_0	Amplitude of heading angle used for zigzag manoeuvre
ω	Frequency of oscillation
∇	Immersed volume

Abbreviations

ACS	Aft Control Surface(s)
AMC	Australian Maritime College, an Institute of The University of Tasmania
ATT	Aft Trim Tank
CFD	Computational Fluid Dynamics
CIS	Cavitation Inception Speed
CLR	Centre of Lateral Resistance
COTS	Commercial Off-The-Shelf
DARPA	Defence Advanced Research Projects Agency (US)
DDD	Deep Dive Depth
DERA	Defence, Evaluation and Research Agency, UK
DES	Detached Eddy Simulation
DGA	Direction Générale de l'Armement, the French Government Defence procurement agency
DRDC	Defence Research and Development Canada
DREA	Defence Research Establishment Atlantic
DST Group	Defence Science and Technology Group, Australia (formerly DSTO)
DSTO	Defence Science and Technology Organisation, Australia (now DST Group)
FCS	Forward Control Surface(s)
FSC	Free Surface Correction
FTT	Forward Trim Tank
GRP	Glass Reinforced Plastic
HPMM	Horizontal Planar Motion Mechanism
IHSS	Iranian Hydrodynamic Series of Submarines
IMO	International Maritime Organisation
ITTC	International Towing Tank Conference
LCB	Position of the Longitudinal Centre of Buoyancy
LCG	Position of the Longitudinal Centre of Gravity
LES	Large Eddy Simulation
MBT	Main Ballast Tank
MDTF	Marine Dynamics Test Facility
MED	Maximum Excursion Depth
MLD	Manoeuvring Limitation Diagram

NACA	National Advisory Committee for Aeronautics
PMB	Parallel Middle Body
PMM	Planar Motion Mechanism
QPC	Quasi-Propulsive Coefficient
RANS	Reynolds Averaged Navier–Stokes
RNLN	Royal Netherlands Navy
rpm	Revolutions Per Minute
SME	Safe Manoeuvring Envelope
SOE	Safe Operating Envelope
SOP	Standard Operating Procedure
SSBN	Nuclear-Powered Ballistic Submarine
SSK	Conventionally Powered Submarine
SSN	Nuclear-Powered Attack Submarine
SSPA	Swedish maritime consulting organisation
VPMM	Vertical Planar Motion Mechanism
WOF	Wake Objective Function

Chapter 1
Introduction

Abstract Submarines are very specialised vehicles, and their design is extremely complex. This book deals with only the hydrodynamics aspects of submarines, and a knowledge of ship hydrodynamics is assumed. The principles of submarine geometry are outlined in this chapter, covering those terms which are not common to naval architecture, such as: axisymmetric hull; sail; aft body; fore body; control surfaces; casing; and propulsor. Over the years a number of different unclassified submarine geometries have been developed to enable organisations to benchmark results of their hydrodynamics studies in the open literature. These geometries have also been used to provide initial input to the design of new submarine shapes. A summary of some of the more widely used geometries is given, along with references to enable the reader to obtain further information as required.

1.1 General

Submarines are very specialised vehicles, and their design is extremely complex. This book deals only with the hydrodynamics aspects of submarines, and a knowledge of ship hydrodynamics is assumed. Readers are referred to texts such as Rawson and Tupper (2001) for information about surface ship concepts.

Although nuclear powered submarines can be much larger than many surface ships, it is traditional to refer to all submarines as "boats" regardless of their size. This convention is retained in this book.

Some details of a range of modern submarines are given in the Appendix.

1.2 Geometry

Submarine geometry is fairly straightforward; however there are various terms used which are not common to naval architecture in general. Firstly the hull is usually based on an axisymmetric body: one which is perfectly symmetrical around its

M. Renilson, *Submarine Hydrodynamics*,
https://doi.org/10.1007/978-3-319-79057-2_1

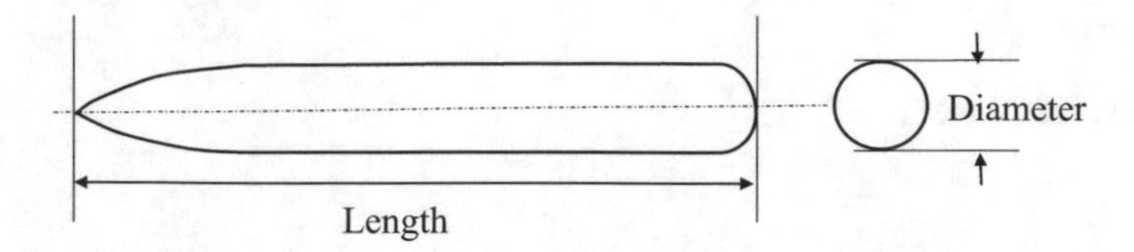

Fig. 1.1 Axisymmetric body

longitudinal axis, as shown in Fig. 1.1. Also indicated in Fig. 1.1 are the length, and the diameter of the axisymmetric body.

For operational purposes it is necessary to add a sail, also known as a bridge fin, to house items such as periscopes, the snorkel and other masts, as shown in Fig. 1.2. This can also be used as a platform to control the boat from when it is on the water surface. For consistency, this will be referred to as the sail, throughout this book. The sail generally has a detrimental effect on the hydrodynamic performance of the submarine.

In addition, forward and aft control surfaces are required to control the boat, as discussed in Chap. 3. Details of the hydrodynamic aspects of the design of these control surfaces are given in Chap. 6. For a boat with a conventional cruciform stern the aft control surfaces will include both an upper and a lower rudder, as shown in Fig. 1.2.

Many modern submarines are propelled by a single propulsor located on the longitudinal axis. This is normally located aft of the aft control surfaces, as shown in Fig. 1.3.

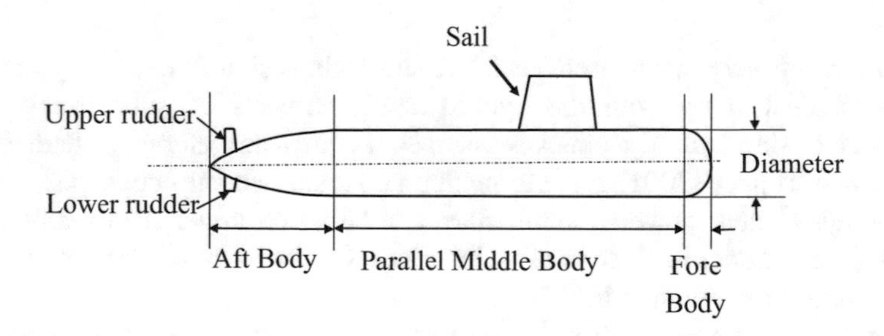

Fig. 1.2 Submarine geometry

Fig. 1.3 Common stern configuration

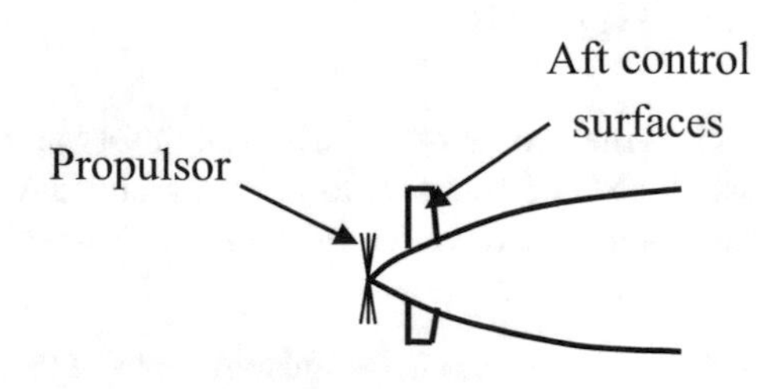

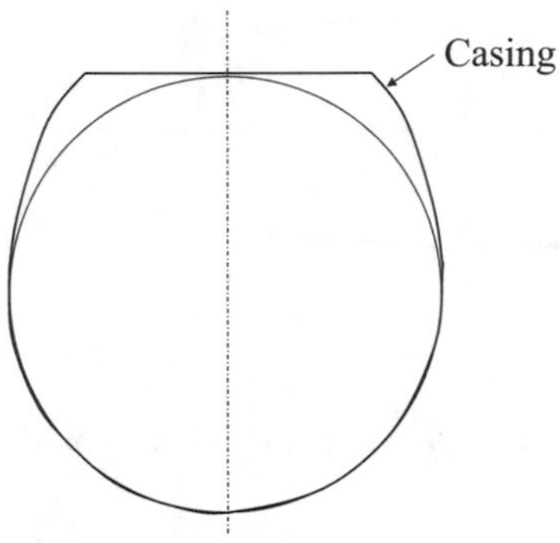

(a) Cross section showing casing

(b) Oberon class submarine on surface (courtesy of Bruce Cartwright)

Fig. 1.4 Casing

Note that the term "propulsor" is often used as this can refer to either a conventional propeller, or a pumpjet, as discussed in Chap. 5.

Although an axisymmetric shape is good for underwater performance it is difficult for the crew to work on the curved upper part of this when the boat is on the surface, and for this reason many submarines are fitted with an external casing, as shown in Fig. 1.4.

In addition to providing a convenient platform to operate from when on the surface, the casing also provides storage space outside the pressure hull which can be useful for operational purposes.

1.3 Standard Submarine Geometries

1.3.1 Series 58

The earliest systematic investigation into the resistance of modern hull forms was conducted in the David Taylor Model Basin and reported in Gertler (1950). This series (Series 58) compressed 24 mathematically related streamlined bodies of revolution with changes in the following geometrical parameters: fineness ratio; prismatic coefficient; nose radius; tail radius; and position of the maximum section. The results are presented in terms of equal volume basis, including estimated appendage resistance due to control surfaces necessary for directional stability.

The shapes of the forms are all defined by a sixth degree polynomial of the form given in Eq. 1.1.

$$r_x = a_1x + a_2x^2 + a_3x^3 + a_4x^4 + a_5x^5 + a_6x^6 \tag{1.1}$$

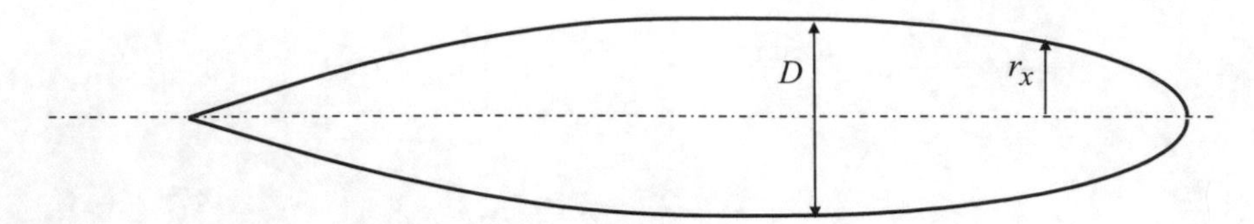

Fig. 1.5 Profile of Series 58 shape

The constants: $a_1; a_2; a_3; a_4; a_5; a_6$ were determined when the values of the geometrical parameters were set (Fig. 1.5).

1.3.2 *Myring Shape*

Myring (1981) developed a standard axi-symmetrical shape suitable for submarine hulls based on an elliptical fore body, a parallel middle body, and a parabolic shaped aft body, as shown in Fig. 1.6.

Using the notation adopted here, the shape is given by Eqs. 1.2–1.4.

Fore body

The shape of the fore body is defined by Eq. 1.2.

$$r_{x_f} = \frac{D}{2}\left[1 - \left(\frac{x_f}{L_F}\right)^2\right]^{\frac{1}{n_f}} \tag{1.2}$$

where, r_{x_f} is the radius of the section at a distance x_f in the x-direction from the rearmost part of the fore body, as shown in Fig. 1.6. L_F is the length of the fore body, D is the hull diameter, and n_f is a coefficient which defines the fullness of the fore body. When $n_f = 2$ the bow profile is an elliptical form.

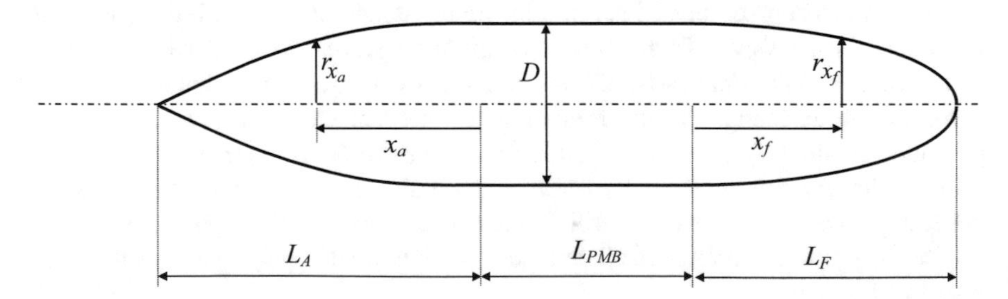

Fig. 1.6 Profile of Myring shape

Parallel middle body

The shape of the parallel middle body is defined by Eq. 1.3.

$$r_{x_{PMB}} = \frac{D}{2} \tag{1.3}$$

where, $r_{x_{PMB}}$ is the radius of the parallel middle body.

Aft body

The shape of the aft body is defined by Eq. 1.4

$$r_{x_a} = \frac{D}{2} - \left[\frac{3D}{2L_A^2} - \frac{\tan\alpha_t}{L_A}\right]x_a^2 + \left[\frac{D}{L_A^3} - \frac{\tan\alpha_t}{L_A^2}\right]x_a^3 \tag{1.4}$$

where, α_t is the half tail cone angle, L_A is the length of the aft body, and x_a is the distance in the *x*-direction aft of the forward most part of the aft body.

1.3.3 DRDC Standard Submarine Model

A standard submarine model was developed for a series of hydrodynamic experiments jointly funded by the DRDC and the RNLN as described by Mackay (2003). This hull form is typical of a SSK configuration. It has subsequently been tested at a number of different facilities, and has also been used for numerous CFD investigations.

A profile of the DRDC standard submarine model is given in Fig. 1.6 taken from Mackay (2003).

The standard submarine model hull is specified in three sections: fore body; parallel middle body; and aft body. The parent (basis) hull has $L/D = 8.75$.

Fore body

The length of the fore body is $1.75D$.

The shape of the fore body is defined by Eq. 1.5.

$$\frac{r_{x_f}}{D} = 0.8685\sqrt{\frac{x_F}{D}} - 0.3978\frac{x_F}{D} + 0.006511\left(\frac{x_F}{D}\right)^2 + 0.005086\left(\frac{x_F}{D}\right)^3 \tag{1.5}$$

where r_{x_f} is the radius of the section at a distance x_F in the *x*-direction from the forward perpendicular measured aft, as shown in Fig. 1.7 and D is the hull diameter.

Parallel middle body

The parallel middle body has a length of $4D$.

The shape of the parallel middle body is defined by Eq. 1.6.

$$r_{x_{PMB}} = \frac{D}{2} \tag{1.6}$$

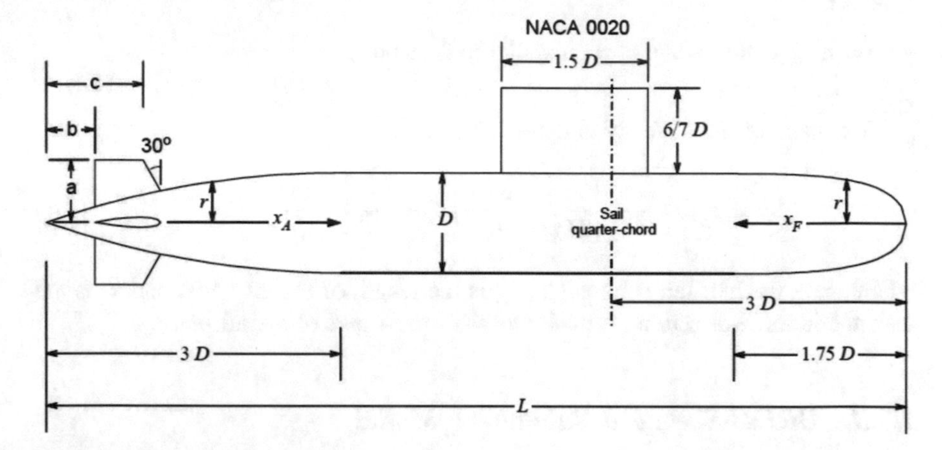

Fig. 1.7 Profile of DRDC standard submarine model (taken from Mackay 2003—not to scale)

where, $r_{x_{PMB}}$ is the radius of the parallel middle body, and D is the diameter.

Aft body

The aft body has a length of 3*D*.

The shape of the aft body is defined by Eq. 1.7

$$\frac{r_{x_A}}{D} = \frac{1}{3}\left(\frac{x_A}{D}\right) - \frac{1}{18}\left(\frac{x_A}{D}\right)^2 \tag{1.7}$$

where, r_{x_A} is the radius of the section at a distance x_A in the *x*-direction from the aft perpendicular measured forward, as shown in Fig. 1.7 and D is the hull diameter.

1.3.4 DARPA Suboff Model

The submarine technology office of DARPA funded a program to assist with the development of submarines, part of which involved the development of a standard submarine hull form, known as "Suboff" (Groves et al. 1989). This hull form is typical of an SSN configuration, and has a notional scale ratio of 1/24, giving a full scale length of 105 m.

Captive model experiments have been conducted using this hull form, with and without appendages, see for example Huang et al. (1989) and Roddy (1990).

This hull form has been widely used since for a number of investigations, including the validation of CFD for manoeuvring coefficients.

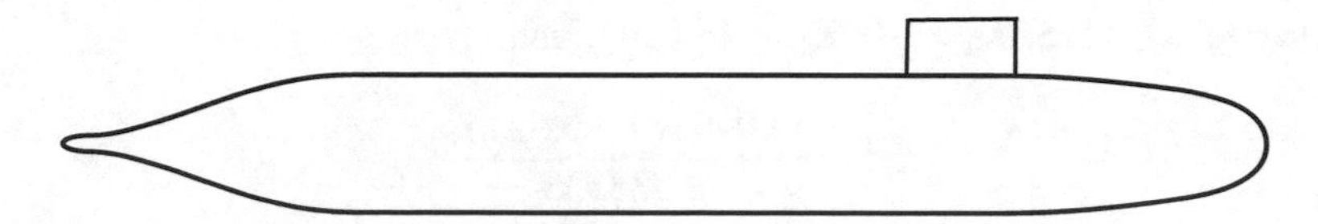

Fig. 1.8 Profile of DARPA Suboff (not to scale)

A profile of Suboff is given in Fig. 1.8.

The Suboff model has an axisymmetric hull with an overall length of 4.356 m and a maximum diameter of 0.508 m.

Fore body

The length of the fore body is $2D$ (1.016 m).

The shape of the fore body is defined by Eq. 1.8.

$$r_{x_f} = \frac{D}{2}\Big[1.126395101x_{SUBOFF}(0.3x_{SUBOFF} - 1)^4 + 0.442874707x_{SUBOFF}^2(0.3x - 1)^4(1.2x + 1)\Big]^{1/2.1} \tag{1.8}$$

<!endaligned >

where r_{x_f} is the radius of the section in feet at a distance x_{SUBOFF} in feet aft of the forward perpendicular, and D is the hull diameter.

Parallel middle body

The length of the parallel middle body is $4.39D$ (2.229 m).

The shape of the parallel middle body is defined by Eq. 1.9.

$$r_{x_{PMB}} = \frac{D}{2} \tag{1.9}$$

where, $r_{x_{PMB}}$ is the radius of the parallel middle body, and D is the diameter.

Aft body

The length of the aft body is $2.19D$ (1.111 m). This comprises a main part which has a length of 1.016 m, and an end cap which has a length of 0.095 m. The aft perpendicular is defined as being at the forward end of the end cap.

The shape of the aft body from the aft end of the parallel middle body to the end cap is defined by Eq. 1.10.

$$r_{x_a} = \frac{D}{2}\left[r_h^2 + r_h K_0 \xi^2 + \left(20 - 20r_h^2 - 4r_h K_0 - \frac{1}{3}K_1\right)\xi^3 + \left(-45 + 45r_h^2 + 6r_h K_0 + K_1\right)\xi^4 + \left(36 - 36r_h^2 - 4r_h K_0 - K_1\right)\xi^5 + \left(-10 + 10r_h^2 + r_h K_0 + \frac{1}{3}K_1\right)\xi^6\right]^{1/2} \tag{1.10}$$

where: $r_h = 0.1175$; $K_0 = 10$; $K_1 = 44.6244$; and:

$$\xi = \frac{13.979167 - x_{SUBOFF}}{3.333333}$$

Sail

Suboff has a sail which could be located on the hull at the top dead centre with its leading edge positioned 0.924 m (1.820*D*) aft of the forward perpendicular, and the trailing edge 1.293 m aft of the forward perpendicular, giving an overall sail chord of 0.368 m (0.724*D*), as shown in Fig. 1.8. The sail is fitted with a sail cap.

Further details of the sail shape can be obtained from Groves et al. (1989).

Stern appendages

There are four identical appendages which could be mounted on the hull at angles of 0°, 90°, 180° and 270°. These could be fitted to the hull at three different longitudinal positions.

In addition, two different ring wings could be fitted to the Suboff.

Further details of the stern appendages can be obtained from Groves et al. (1989).

1.3.5 Iranian Hydrodynamic Series of Submarines (IHSS)

The Iranian Hydrodynamic Series of Submarines was developed specifically to serve as a basis for systematically investigating the hydrodynamics of modern submarines (Moonesun and Korol 2017). It has an elliptical fore body, a parallel middle body, and a conical aft body, with no propeller. It has a sail with a symmetrical NACA foil section. A diagram of the IHSS standard hull form is given in Fig. 1.9.

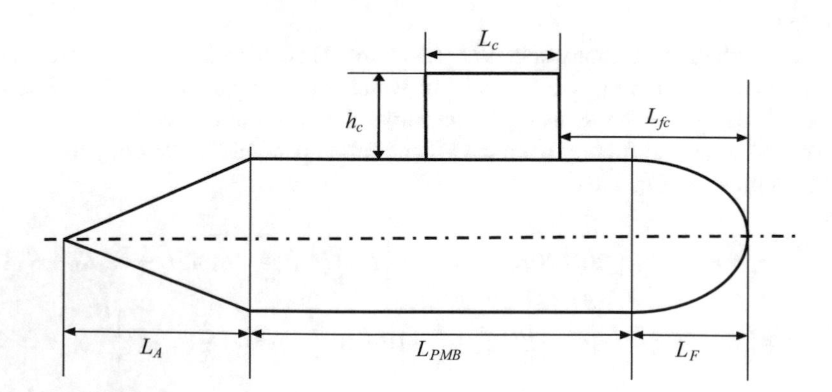

Fig. 1.9 Definition of parameters in IHSS (Moonesun and Korol 2017)

Table 1.1 Definition of decoding IHSS 15 digit code

Digits	Meaning	Digits in example	Value in example
1st three	Ratio of length to diameter	105	10.5
4th and 5th	Ratio of fore body length to overall length	25	0.25
6th and 7th	Ratio of parallel middle body to overall length	70	0.70
8th and 9th	Ratio of distance of leading edge of sail from bow to overall length	35	0.35
10th and 11th	Ratio of length of sail to overall length	17	0.17
12th and 13th	Ratio of height of sail to diameter	80	0.80
14th and 15th	Sail foil shape	25	NACA 0025

A hull form within the IHSS is specified using a 15 digit code with the first seven digits defining the main hull, and the remaining eight digits the sail.

Thus, code IHSS 1052570-35178025 is described in Table 1.1.

1.3.6 Joubert/BB1/BB2

A concept design of a large SSK was carried out for the Australian Department of Defence by Joubert (2004, 2006). This has been used by a number of organisations as a standard submarine hull form, and is referred to as BB1. BB1 was subsequently modified with changes to the aft control surfaces and sail as described in Overpelt et al. (2015). The modified version is known as BB2, and has been the subject of considerable further work, including: Bettle (2014), Carrica et al. (2016) and Pook et al. (2017). The full scale principal particulars are given in Table 1.2, and profiles are given in Fig. 1.10.

Table 1.2 Principal particulars for BB1 and BB2

Dimension	Full scale value (m)
Length overall	70.2
Beam	9.6
Depth (to deck)	10.6
Depth (to top of sail)	16.2
Propeller diameter	5.0

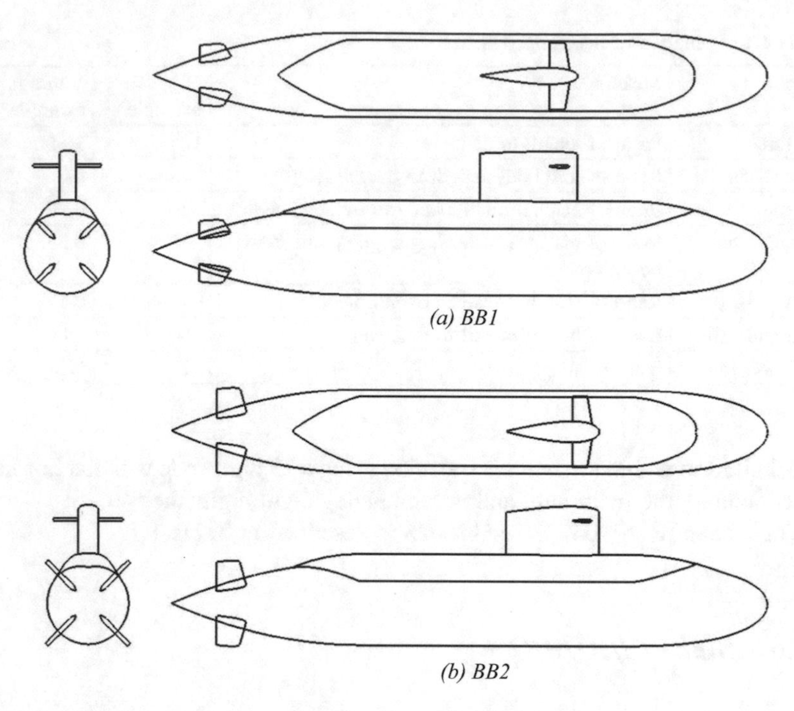

Fig. 1.10 Drawings of BB1 and BB2 (courtesy of DST Group)

References

Bettle MC (2014) Validating design methods for sizing submarine tailfins. In: Proceedings of warship 2014, Bath, UK, 18–19 June 2014

Carrica PM, Kerkvliet M, Quadvlieg F, Pontarelli M, Martin E (2016) CFD simulations and experiments of a manoeuvring generic submarine and prognosis for simulation of near surface operation. In: Proceedings of the 31st symposium on naval hydrodynamics, Monterey, CA, USA, 11–16 Sept 2016

Gertler M (1950) Resistance experiments on a systematic series of streamlined bodies of revolution—for application to the design of high-speed submarines. David W Taylor Model Basin Report C-297, Apr 1950

Groves NC, Huang TT, Chang MS (1989) Geometric characteristics of the DARPA SUBOFF model. David Taylor Research Centre, 1989

Huang TT, Liu HL, Groves NC (1989) Experiments of the DARPA SUBOFF program, DTRC/SHD-1298-02, Dec 1989

Joubert PJ (2004) Some aspects of submarine design, part 1. Hydrodynamics, defence science and technology organisation. Australian Government, Department of Defence, DSTO-TR-1622, Oct 2004

Joubert PJ (2006) Some aspects of submarine design, part 2. Shape of a submarine 2026. Defence Science and Technology Organisation, Australian Government, Department of Defence, DSTO-TR-1920, Dec 2006

Mackay M (2003) The standard submarine model: a survey of static hydrodynamic experiments and semiempirical predictions. Defence R&D Canada, June 2003

Moonesun M, Korol Y (2017) Naval Submarine body form design and hydrodynamics. LAP LAMBERT Academic Publishing. ISBN: 978-620-2-00425-1

Myring DF (1981) A theoretical study of the effects of body shape and mach number on the drag of bodies of revolution in subcritical axisymmetric flow. Royal Aircraft Establishment Technical Report 81005, Jan 1981

Overpelt B Nienhuis B, Anderson B (2015) Free running manoeuvring model tests on a modern generic SSK class submarine (BB2). In: Proceedings of Pacific 2015, Sydney, Australia, 6–8 Oct 2015

Pook DA Seil G, Nguyen M, Ranmuthugala D, Renilson MR (2017) The effect of aft control surface deflection at angles of drift and angles of attack. In Proceedings of warship 2017: naval submarines and UUVs. Royal Institution of Naval Architects, Bath, UK

Rawson KJ, Tupper EC (2001) Basic ship theory, 5th edn. Butterworth-Heinemann

Roddy RF (1990) Investigation of the stability and control characteristics of several configurations of the DARPA SUBOFF model (DTRC Model 5470) From captive model experiments, DTRC/SHD-1298-08 Sept 1990

Chapter 2
Hydrostatics and Control

Abstract A submarine must conform to Archimedes' Principle, which states that a body immersed in a fluid has an upward force on it (buoyancy) equal to the weight of the displaced fluid, (displacement). There are two different definitions of submerged displacement: one that doesn't include the mass of fluid in the free flooding spaces (hydrostatic displacement), which is used by submarine naval architects, and one that does include the mass of the fluid in the free flooding spaces (form displacement), which is used by submarine hydrodynamicists. For equilibrium in the vertical plane the mass must be balanced *exactly* by the buoyancy force. As compressibility affects the buoyancy, it is not possible for a submarine to be in stable equilibrium in the vertical plane. Submarines are fitted with ballast tanks to enable the mass to be changed. Ballast tanks fit into two categories: those used for major adjustment of mass (main ballast tanks); and those used for minor adjustments (trim tanks). The effect of each tank is plotted and this is compared with the changes in mass and trimming moment possible during operations using a trim polygon to determine whether the ballast tanks are adequate. Transverse stability of a submarine is discussed, including particular issues that arise when passing through the free surface, when on the seabed, or when surfacing through ice. On the water surface, metacentric height (GM) is important, whereas below the surface it is the distance between the centre of buoyancy and the centre of gravity (BG) which governs the transverse stability of a submarine. Various transverse stability criteria are presented, both for surfaced and submerged submarines.

2.1 Hydrostatics and Displacement

As with any object in a fluid, a submarine must conform to Archimedes' Principle, which states that a body immersed in a fluid has an upward force on it (buoyancy) equal to the weight of the displaced fluid, (displacement). This applies whether the submarine is floating on the water surface, or deeply submerged. Readers are referred to texts such as Rawson and Tupper (2001), or Renilson (2016), for general information about ship stability and hydrostatics.

M. Renilson, *Submarine Hydrodynamics*,
https://doi.org/10.1007/978-3-319-79057-2_2

When it is floating on the water surface, less of the boat is under the water, and hence the buoyancy and the displacement will be less than when it is submerged.

A key feature of a submarine is its ability to vary its mass, and hence to change from floating on the water surface, to being fully submerged, and vice versa. Therefore, a submarine will have a submerged displacement, for when it is operating under the surface, and a surface displacement for operations on the water surface. It is quite normal to have more than one surfaced displacement, depending on the level of reserve buoyancy required for any given operation. This principle is exactly the same as that for a conventional vessel, which may operate at more than one draught.

In addition, there are two different definitions of submerged displacement as given in Table 2.1.

Hydrostatic displacement is usually used by naval architects when considering the mass and buoyancy balance of the submarine, particularly at the design stage. The free flood water, such as that in the main ballast tanks and under casings, is excluded, as this can be considered to be irrelevant to either the total mass of the vessel, or its total buoyancy.

On the other hand, the form displacement is usually used by hydrodynamicists, who are concerned with the mass of the submarine which needs to be propelled, and manoeuvred. In this case, as the mass of the water in the main ballast tanks and under casings needs to move with the submarine, it is necessary that it be considered.

These two definitions of displacement will each have a centre of buoyancy and a centre of gravity which are different to each other, as given in Table 2.2 and Fig. 2.1.

It is obviously very important to ensure that it is clearly understood which definition of displacement is being used!

Table 2.1 Definitions of submerged displacement

Definition	Symbol	Description
Hydrostatic displacement	Δ_H	Total mass, other than free flood water
Form displacement	Δ_F	Total mass, including free flood water

Table 2.2 Centres of gravity and buoyancy

Definition	Centre of gravity	Centre of buoyancy
Hydrostatic displacement	G_H	B_H
Form displacement	G_F	B_F

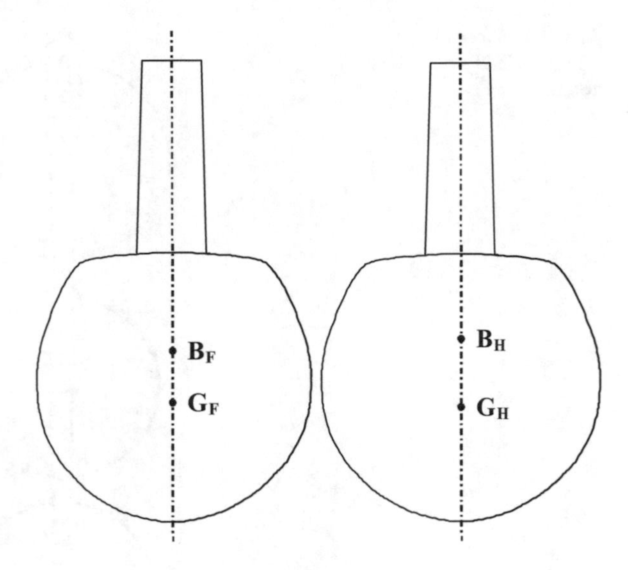

Fig. 2.1 Centres of gravity and buoyancy

As the righting moment at an angle of heel must be the same for these two definitions of displacement, the relationship between the centres can be obtained from Eq. 2.1.

$$B_F G_F \times \Delta_F = B_H G_H \times \Delta_H \tag{2.1}$$

Therefore:

$$\frac{B_H G_H}{B_F G_F} = \frac{\Delta_F}{\Delta_H} \tag{2.2}$$

2.2 Static Control

2.2.1 Control in the Vertical Plane

The downward force due to the mass multiplied by gravity must be balanced by the upward buoyancy force given by the immersed volume multiplied by the water density and gravity.

Unlike a surface ship, in the case of a deeply submerged submarine the immersed volume cannot be increased by increasing the vessel's draught. Thus, for equilibrium in the vertical plane the mass must be balanced *exactly* by the buoyancy force. Clearly this is difficult, if not impossible to achieve.

To further complicate the issue, the deeper the submarine is operating, the greater the water pressure acting on it will be, resulting in the hull being compressed. This will reduce the immersed volume, and hence the upward buoyancy

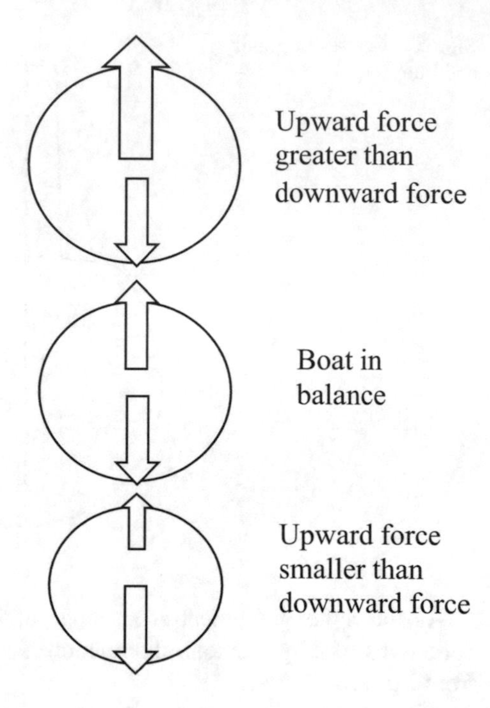

Fig. 2.2 Effect of compressibility on buoyancy force

force. Conversely, if the submarine moves closer to the surface the water pressure acting on it will be less, and hence the immersed volume and the upward buoyancy force will be greater. This is illustrated in Fig. 2.2.

The magnitude of this compressibility effect will depend on the submarine structure, however it is important to recognise that many modern submarines are fitted with acoustic tiles, which themselves are compressible, increasing the magnitude of this problem.

Thus, the best that can be achieved is for a submarine to be in unstable equilibrium at a given depth of submergence. A slight upward or downward movement from this position will result in the boat moving away from this initial position.

Further, small changes in sea water density occur in the vertical and horizontal planes, particularly close to coasts. These will also have a significant influence on the ability to control the submarine in the vertical plane.

In addition, the mass on board will change during a voyage due to use of consumables and/or discharge of weapons.

Hence, it is necessary to have the ability to make small changes in mass of the boat very quickly, which is done by a series of ballast tanks, as discussed in Sect. 2.3. Even then, it is very difficult to control a submarine in the vertical plane at zero forward speed, and so it is necessary to make use of hydrodynamic forces, as discussed in Chap. 3.

2.2.2 Transverse Stability

For a submerged submarine to be stable in roll, known as transverse stability, the centre of buoyancy must be above the centre of gravity, as shown in Fig. 2.3. In this case, if the boat is heeled to a small angle, as shown in Fig. 2.4, the hydrostatic moment on it will cause it to return to the upright. On the other hand, if the centre of gravity is above the centre of buoyancy, and an external moment causes it to be heeled to a small angle, then the hydrostatic moment will cause it to continue to heel, as shown in Fig. 2.5.

The measure of transverse stability is then given by the distance BG. As noted in Sect. 2.1, for a given submarine this distance will be different depending on whether it is the hydrostatic or form displacement which is being considered. A positive BG value (B above G) is necessary for a submerged submarine, and is usually easy to achieve, since in many ways the more critical element of transverse stability occurs when surfacing or submerging, as discussed in Sect. 2.5.

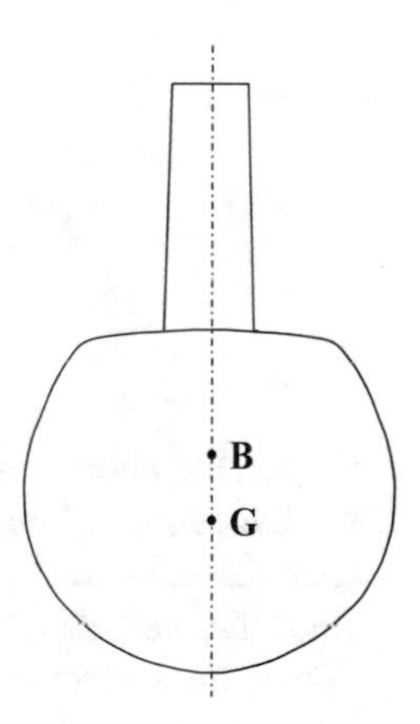

Fig. 2.3 Submerged submarine in stable transverse equilibrium

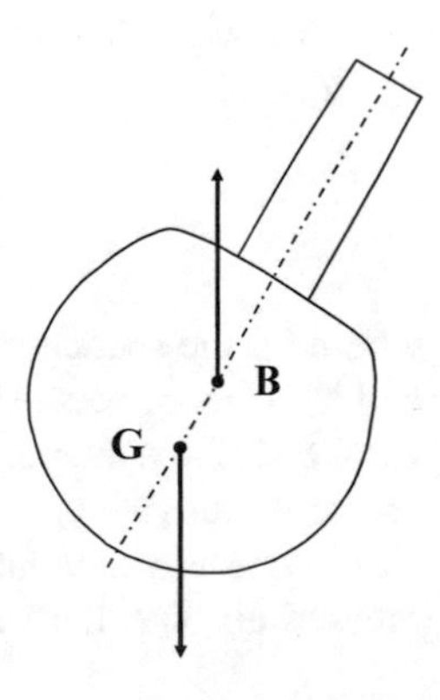

Fig. 2.4 Submerged submarine with small heel angle when B is above G

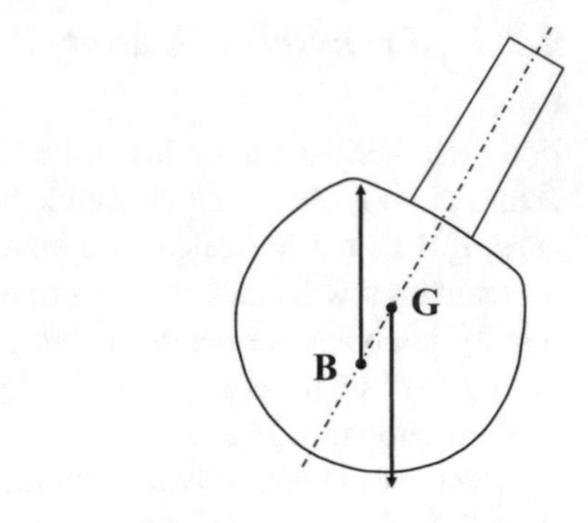

Fig. 2.5 Submerged submarine with small heel angle when G is above B

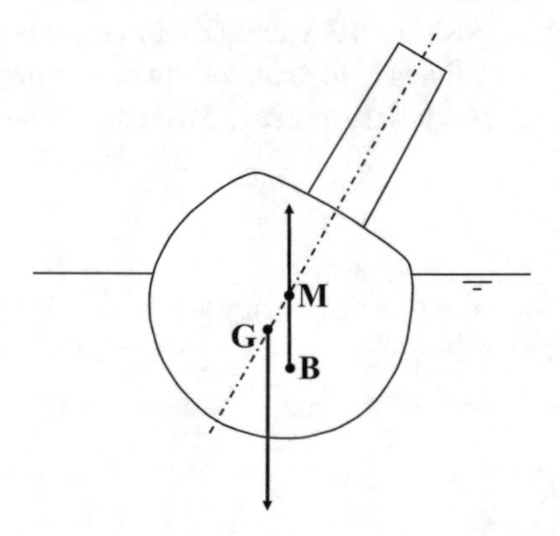

Fig. 2.6 Submarine with a small heel angle floating on the surface

When the submarine is floating on the surface the situation is different. In this case, the centre of buoyancy moves transversely when the boat heels. For small angles the upward force through the centre of buoyancy always acts through the metacentre, designated as "M" in Fig. 2.6.

Thus, for a surfaced boat to be in stable equilibrium when upright, the position of the metacentre has to be above the centre of gravity, and the measure of the stability is given by the distance GM.

The vertical distance between the centre of buoyancy and the metacentre is given by Eq. 2.3.

$$BM = \frac{I}{\nabla} \tag{2.3}$$

where I is the second moment of area of the waterplane around the longitudinal axis and ∇ is the immersed volume. When the submarine is submerged I will be equal to zero, and hence the position of the metacentre will be the same as the position of the centre of buoyancy.

If there are any fluids on board the submarine in tanks which are not fully pressed up then the centre of gravity of these fluids will also move transversely

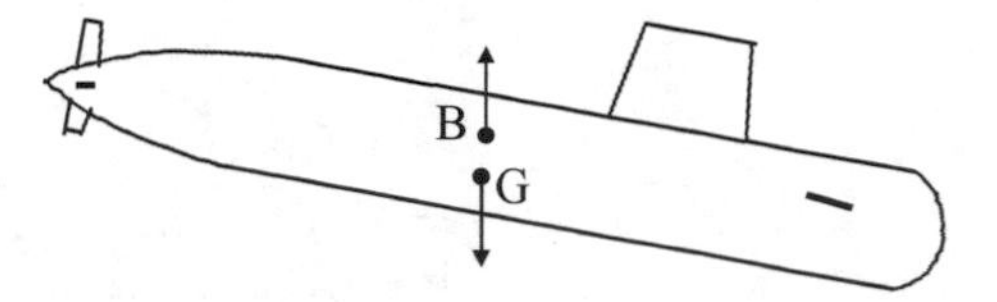

Fig. 2.7 Submerged submarine at angle of trim

when the submarine heels. This can be considered as a raising in the position of the centre of gravity from G, to G_F. Note that the subscript "F" in this case does not refer to "form" as discussed in Sect. 2.1, but to the position of the centre of gravity corrected due to free surface.

The vertical distance between the centre of gravity and the centre of gravity corrected for free surface, is known as the Free Surface Correction (FSC). This is dependent on the second moment of area of the fluid in the various tanks, and the density of the fluid in the tanks, not directly on the mass of the fluid in the tanks.

2.2.3 Longitudinal Stability

As with transverse stability, the same principles apply to a submerged submarine as to a floating surface ship, however the lack of a waterplane results in a very small restoring moment in the longitudinal direction if the submarine is trimmed, as shown in Fig. 2.7.

Thus, it is essential to have the longitudinal position of the centre of gravity lined up with the longitudinal position of the centre of buoyancy. As the longitudinal position of the centre of gravity moves during a voyage due to use of consumables, firing of weapons, etc., it is necessary to be able to adjust this by use of ballast tanks, as discussed in Sect. 2.3.

2.3 Ballast Tanks

2.3.1 Categories of Ballast Tanks

Ballast tanks fit into two different categories:

(a) those used for major adjustment of submarine mass to allow it to operate on the surface or submerged (main ballast tanks); and
(b) those used for minor adjustments to keep the submarine balanced when submerged (trim and compensation system).

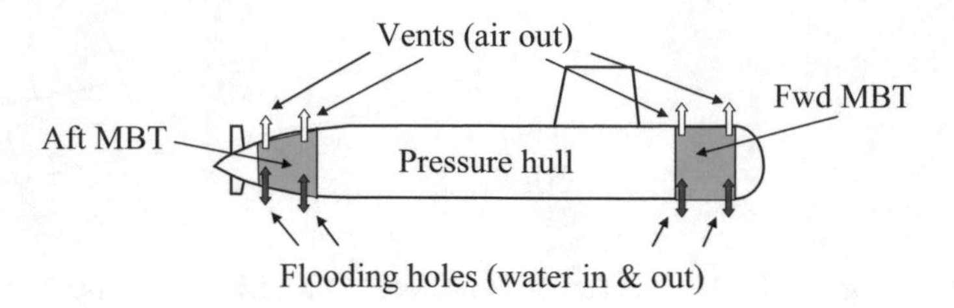

Fig. 2.8 Schematic of typical main ballast tank system

2.3.2 *Main Ballast Tanks*

The Main Ballast Tanks (MBTs), are usually ballast tanks external to the pressure hull, which are free flooding when the submarine is submerged, as shown in Fig. 2.8.

The purpose of the MBTs is to allow major adjustment of the submarine mass to enable it to operate submerged as well as on the water surface. Water and air enter and leave the MBTs through flooding holes at the bottom and vents at the top of the tanks.

When the submarine is on the water surface the MBTs are flooded by opening the vent valves, allowing water to enter the MBTs through the flooding holes. The size of the vents and the flooding holes will have a direct effect on the length of time that it takes for the MBTs to fill, and hence on how long it will take for the submarine to submerge. Ideally the size should be chosen such that all the tanks flood at the same time.

The size of the flooding holes will also affect the hydro-acoustic signature when the submarine is operating submerged, as they cause a disturbance to the flow around them. Small flooding holes may cause problems with over-pressure, and stability issues on the surface.

2.3.3 *Trim and Compensation Ballast Tanks*

During operations the mass and longitudinal centre of gravity of a submarine will change due to use of consumables including fuel, and weapons discharge. In addition, changes in seawater density, hull compressibility and surface suction when operating close to the surface will all result in the need to be able to make small changes to the submarine mass and longitudinal centre of gravity.

The trim and compensation ballast tanks are used to make these small adjustments. A schematic of such a typical system is shown in Fig. 2.9.

Ideally the compensating tanks should be close to the longitudinal centre of gravity, whilst the trim tanks should be at the extremities of the submarine. In addition, tanks specifically designed to compensate for weapons discharge should

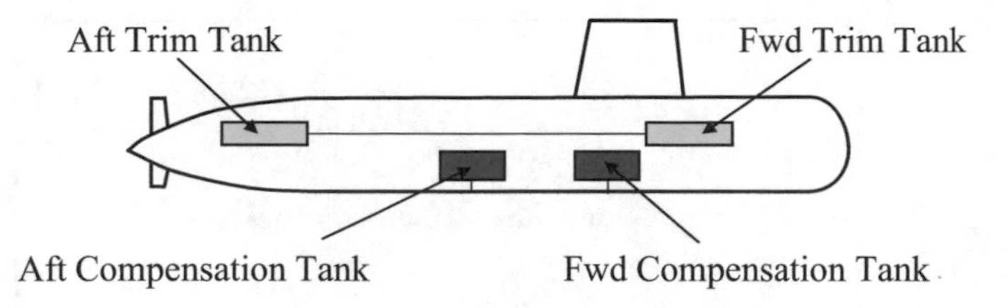

Fig. 2.9 Schematic of typical trim and compensation ballast tanks

be located as close as possible to the appropriate location. Some conventional submarines also have quick dive tanks forward which can be flooded rapidly to assist the boat to leave the surface quickly. These are then emptied once the submarine has submerged.

Trim and compensation tanks can be either hard tanks, which are fully exposed to the external water pressure, or soft tanks which are not. Tanks which are used to compensate for changes in mass are required to be hard, and their systems have to be designed with great care to be able to survive the deep diving depth. A credible failure is often considered to be an uncontrolled leak in such a system, and Standard Operating Procedures (SOPs) need to be developed for such an occurrence, which will also influence the Safe Operating Envelopes, as discussed in Sect. 3.10. Tanks used only to adjust the longitudinal centre of gravity can be soft tanks, meaning that these can be much lighter as their structure does not have to withstand the deep diving depth.

2.4 Trim Polygon

At the design stage it is necessary to determine whether the trim and compensation ballast tanks are adequate to cope with all possible changes in submarine mass and longitudinal centre of gravity. To do this, the effect of each tank is plotted as a function of mass and trimming moment as shown in Fig. 2.10.

A schematic of the ballast tank conditions for the various points on the polygon is given in Fig. 2.11.

In Figs. 2.10 and 2.11 the position indicated by A represents the case where all the tanks are empty. Position B is where the forward trim tank only is filled. As can be seen, this represents an increase in mass (assuming that the forward tank can be filled from external to the submarine) and a forward trimming moment.

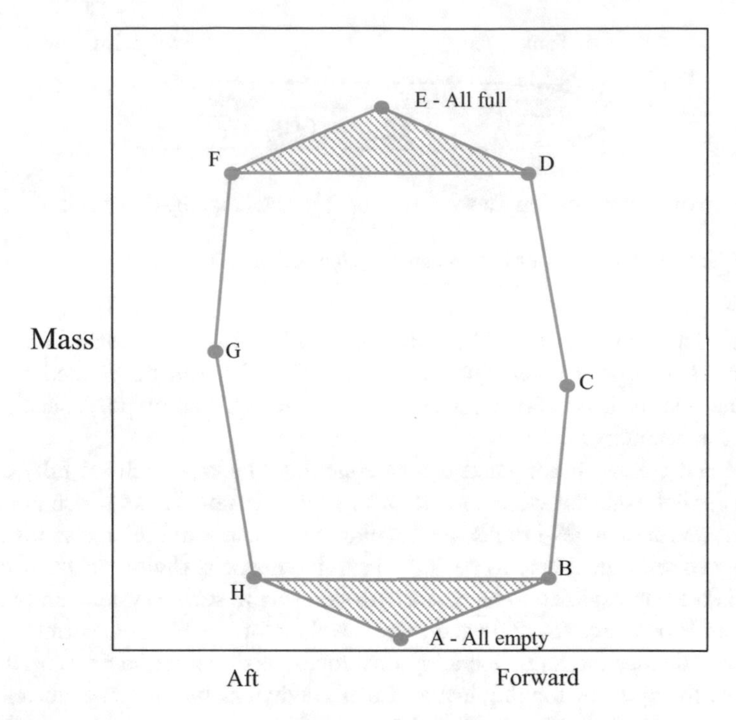

Fig. 2.10 Polygon showing the effect of trim and compensation ballast tanks (see Fig. 2.11 for schematic of ballast tank conditions)

Position C represents the condition where the forward compensation tank is also filled. The effect of this will be an increase in mass. As this is slightly forward of the longitudinal centre of gravity there will also be a small forward trimming moment as shown.

Position D represents the case where both compensation tanks are filled.

Position E represents the case where all the tanks are filled. Position F is where the forward trim tank has been emptied, and position G and H are when the forward and aft compensation tanks are emptied respectively.

Note that if the trim tanks are soft tanks, which cannot be filled or emptied from outside the submarine, then it is not possible to ballast the submarine in positions A and E in Figs. 2.10 and 2.11. When the trim tanks are soft tanks the area shaded in orange is not available.

Additional issues such as compressibility can be included, however these are outside the scope of this book.

The resulting polygon indicates the maximum effect that can be achieved by the trim and compensation ballast tank system.

A similar polygon is then prepared to represent all the possible changes in mass and trimming moment due to use of consumables, including fuel, and weapons discharge. The effects of compressibility and surface suction can also be

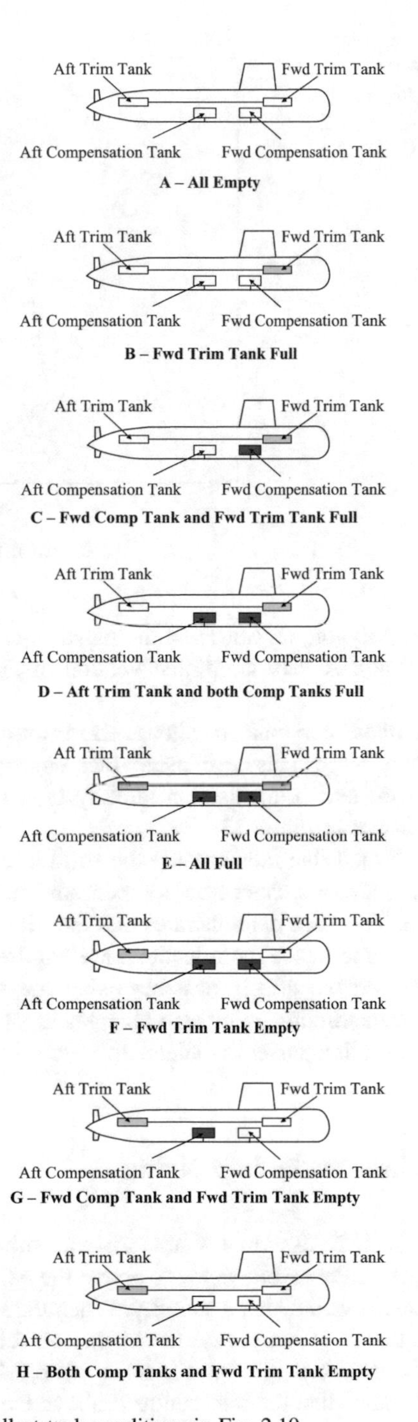

Fig. 2.11 Schematic of ballast tank conditions in Fig. 2.10

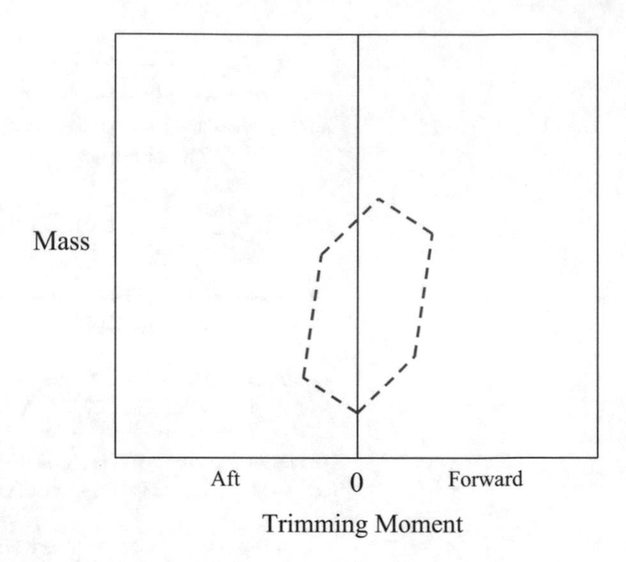

Fig. 2.12 Polygon showing the effect of changes in mass

incorporated into this polygon, as can the anticipated in-service growth (accumulation of mass over time). A very simplified version of the polygon is shown in Fig. 2.12

To ensure that the maximum change in mass and trimming moment that can be caused by factors such as changes in consumables etc. can be adequately compensated for by the trim and compensation tank system, these two polygons are plotted together, as shown in Fig. 2.13.

If any part of the dashed line falls outside the solid line then it is possible that changes to the submarine mass cannot be compensated for by the trim and compensation tank system. This then demonstrates that the trim and compensation tank system is not adequate, and hence modifications are required. In Fig. 2.13 there is sufficient margin between the effects possible using the trim and compensation ballast system, and the maximum anticipated changes in submarine mass and trim, so the trim and compensation system is adequate.

2.5 Stability When Surfacing/Diving

As discussed in sub Sect. 2.2.2, when a submarine is submerged, transverse stability is achieved if the centre of buoyancy is above the centre of gravity, and this distance, BG, is a measure of the boat's stability. When the submarine is floating on the water surface the centre of buoyancy moves transversely as a function of heel angle. For small angles this acts through the metacentre, M. Thus, the measure of stability is then the distance that the metacentre is above the centre of gravity, GM.

As a submarine transitions from floating on the water surface to fully submerged its vertical centres of both buoyancy and gravity will vary, due to a change in

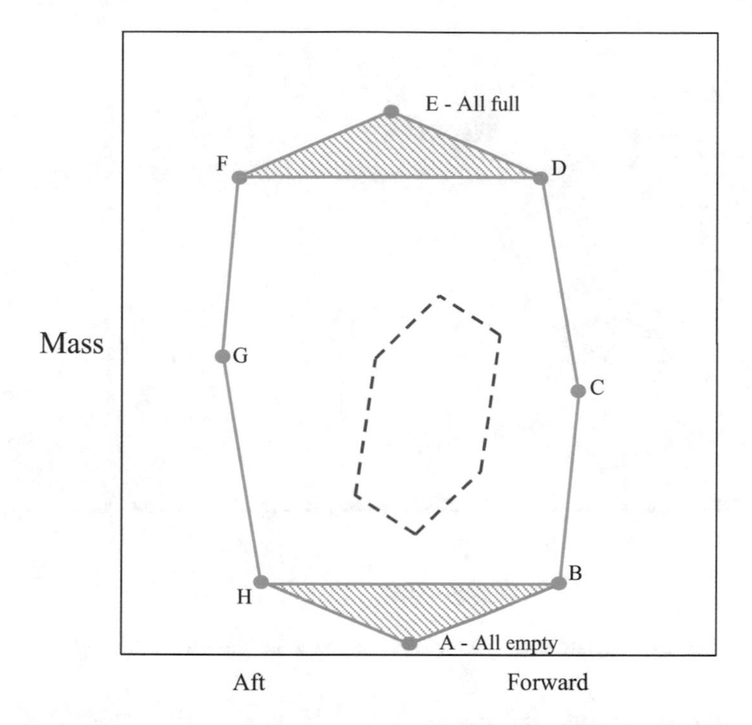

Fig. 2.13 Schematic of trim polygon

immersion of the hull, and the change in mass in the ballast tanks. In addition, the second moment of the waterplane, I, will vary during this process. This will also significantly affect the position of the metacentre, M, Eq. 2.3.

To ensure that a submarine remains stable as it is transiting from being on the surface to fully submerged, a plot of the positions of the various centres is made as a function of draught. An example of this is given in Fig. 2.14. The values of KB, BM, KM, and KG_F (KG corrected for free surface) are given as functions of draught. In this case the lightest surface draught is 4.5 m. The casing is fully submerged at a draught of approximately 5.5 m, and the submarine is fully submerged at a draught of 7 m.

As can be seen, in this case the surface G_FM value is positive, the minimum value of BG_F is positive, and the fully submerged value of BG_F is positive.

An additional complexity, not shown directly in Fig. 2.14, is that, when surfacing, water is retained in the casing, and other free flood spaces, for a period before it can escape. This will raise the centre of gravity above that assumed for the steady state calculations used to generate Fig. 2.14. The length of time that this water takes to escape will depend on the size of the free flooding holes, however as noted in sub Sect. 2.3.2 large holes may affect the hydro-acoustic noise generated when submerged.

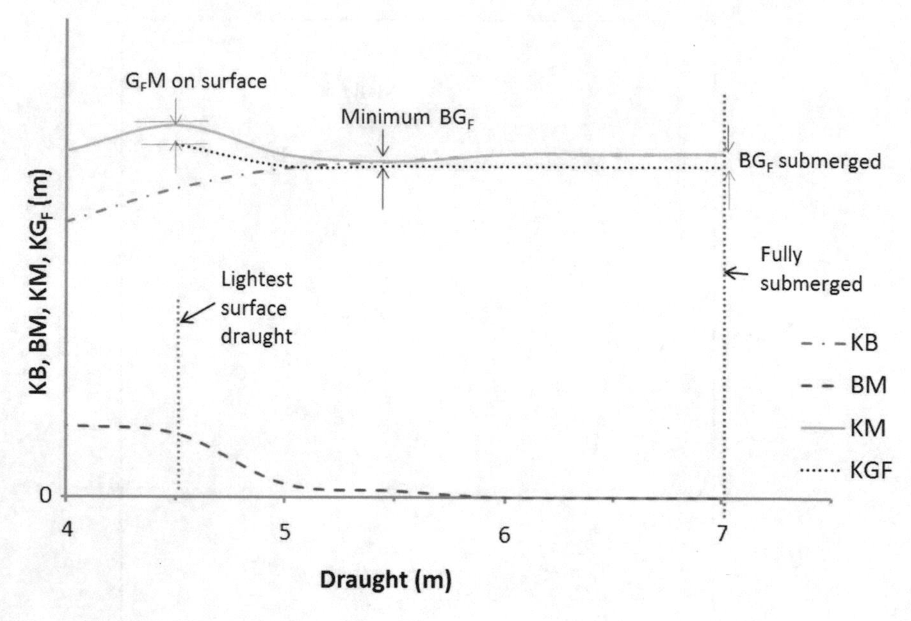

Fig. 2.14 Curve of stability when passing through the water surface

2.6 Stability When Bottoming

From time to time some submarines will sit on the seabed for operational reasons. When this occurs the transverse stability is affected by the upward force on the keel, as shown in Fig. 2.15.

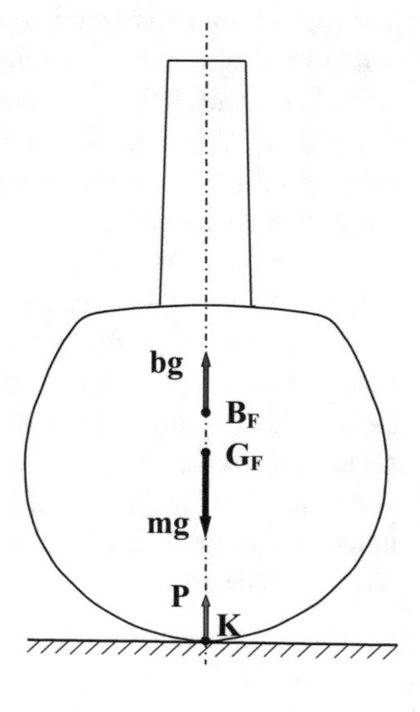

Fig. 2.15 Submarine sitting on the seabed

In Fig. 2.15 the downward mass force, mg, has been increased due to the increase in the mass of the water in the compensating tanks. P is the upward force acting at the keel, K, and the upward force at the centre of buoyancy, bg, is unchanged. Note that the position of G_F will have changed due to the additional water in the compensation tanks. The vertical movement of G_F as a function of the quantity of water in these tanks must be known.

Note also that the form definition of displacement (see Sect. 2.1) is used here. The same outcome would occur if the hydrostatic definition was used instead, however the form definition is used, as it is the internal tanks which have the additional mass of water.

The magnitude of the upward force on the keel can be obtained from equilibrium, Eq. 2.4, where the sign convention of positive downwards has been maintained.

$$mg - bg - P = 0 \tag{2.4}$$

For a small angle of heel of ϕ to starboard (positive) the heeling moments about the keel, K, are given in Eq. 2.5. Again, the standard sign convention of positive being a heeling moment in the clockwise direction is used.

$$\text{Heeling moment } = KG_F \, mg \sin\phi - KB_F \, bg \sin\phi \tag{2.5}$$

For the submarine to return to the upright an anticlockwise moment is required, meaning that the heeling moment must be negative. Thus, for stability, $KB_F \times bg$ must be greater than $KG_F \times mg$. This is important, as it will dictate the maximum amount of water that can be added to the compensating tanks whilst remaining stable. If these are located low in the boat, then there may not be a limit to the amount of water without affecting stability, however if these are high, then it may be necessary to set a limit of the amount of water in the compensating tanks in this condition, to maintain transverse stability.

2.7 Stability When Surfacing Through Ice

Submarines operating under ice must be able to surface by breaking through the ice. The normal procedure is to stop the submarine under thin ice, and then to slowly surface at zero forward speed.

When the sail first makes contact with the ice there will be a downward force from the ice, P, which will increase as ballast is removed, and buoyancy increased, until the ice breaks. This will influence the stability of the submarine, as shown in Fig. 2.16.

The analysis is analogous to the case when the submarine is sitting on the seabed, as discussed in Sect. 2.6, and the maximum force which can be applied can

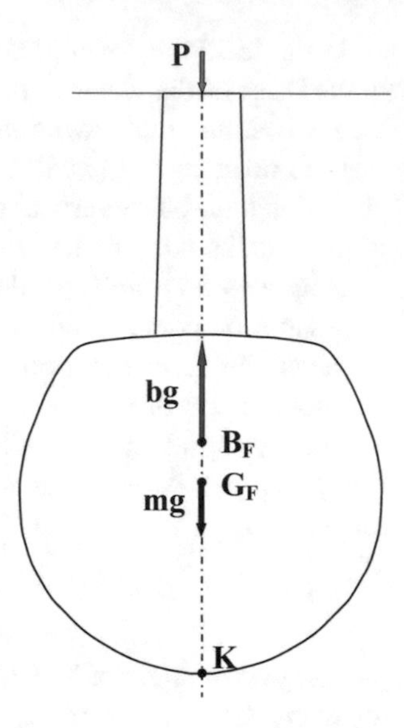

Fig. 2.16 Submarine breaking through ice

be obtained in a similar manner. The initial value of BG must be sufficiently high to allow for the reduction in stability caused by the force at the top of the sail required to break through the ice.

2.8 Stability Criteria

2.8.1 Introduction

A number of organisations, including the IMO have developed stability criteria for surface ships. These are typically based on static criteria ensuring sufficient reserve of stability by using the righting lever curve.

For example, one internationally accepted set of criteria is shown in Table 2.3. This is taken from IMO Resolution MSC.267, which applies to all merchant ships covered by the IMO (2008).

In addition, the IMO recommends use of the so called "weather criterion", which is designed to test the ability of the ship to withstand the combined effects of beam wind and rolling. This is illustrated in Fig. 2.17. In this figure:

1. The ship is subjected to a steady wind pressure acting perpendicular to the ship's centreline which results in a steady wind heeling lever (l_{w1}).
2. From the resultant angle of equilibrium (θ_0), the ship is assumed to roll owing to wave action to an angle of roll (θ_1) to windward. Attention should be paid to the effect of steady wind so that excessive resultant angles of heel are avoided.

3. The ship is then subjected to a gust wind pressure which results in a gust wind heeling lever (l_{w2}).
4. Under these circumstances, area b should be equal to or greater than area a.

The angles of heel are defined as follows:

θ_0 is the angle of heel under the action of a steady wind;
θ_1 is the angle of roll to windward due to wave action;
θ_2 is the angle of downflooding (θ_F) or 50°, whichever is less;
θ_c is the angle of the second intercept between the wind heeling lever (l_w) and the righting lever.

Table 2.3 IMO stability criteria (from IMO 2008)

	Description	Minimum value
1	Area under righting lever curve up to an angle of heel of 30°	0.055 m-rad
2	Area under righting lever curve up to an angle of heel of 40° or the angle of downflooding, if this is less than 40°	0.09 m-rad
3	Area under the righting lever curve between the angles of heel of 30° and 40° or between the angles of heel of 30° and the downflooding angle if this is less than 40°	0.03 m-rad
4	Value of righting lever at an angle of heel equal to, or greater than, 30°	0.2 m
5	Angle of heel for maximum righting lever	Preferably 30°, but not less than 25°
6	Initial metacentric height	0.15 m

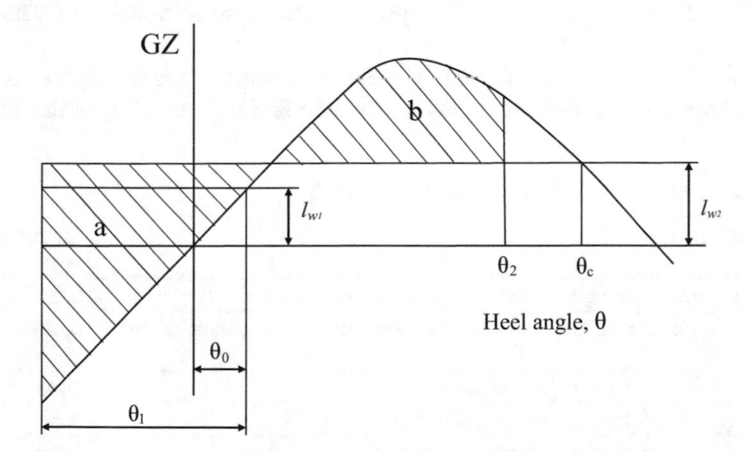

Fig. 2.17 Illustration of the IMO weather criteria (adapted from IMO 2008)

Resolution A.749 gives the methods of calculating these values, and the latest version of this should be referred to if this is to be applied in practice.

Note that various administrations have introduced further criteria, particularly for specialist vessels, or those in restricted service. In addition, military authorities have adopted their own stability criteria, often based on the same, or similar principles to those given above.

These stability criteria are based on empirical data, and may not be directly applicable to the stability of a surfaced submarine. Thus, a number of organisations are using dynamic motions codes to investigate the behaviour of a surfaced submarine in a seaway, and hence develop criteria specifically to address submarine stability. For example Crossland et al. (2017) discusses stability of submarines on the surface, however, criterion values are not given. These are likely to remain classified.

Various classification societies have developed rules for the classification of naval submarines, and these incorporate stability criteria deemed appropriate by the relevant society. Information on some of these criteria are given below, however it should be noted that these are currently in development, and readers are recommended to refer to the most recent classification society rules, as appropriate.

2.8.2 *Bureau Veritas Criteria*

Criteria for transverse stability have been developed for submarines and are given in Bureau Veritas (2016). These are based on US Navy rules developed in the 1970s.

When submerged it is recommended that BG be greater than 0.2 m.

When surfaced the criteria given in Table 2.4 must be met.

Additional criteria to account for dynamic effects are also included. These are summarized in Table 2.5.

In Table 2.5 area A1 is the area between the wind heeling lever curve and the righting lever curve above the angle of static equilibrium, up to 90° or the angle of

Table 2.4 Stability criteria (from Bureau Veritas 2016)

	Description	Minimum value
1	Area under righting lever curve up to an angle of heel of 30°	0.027 m rad
2	Area under the righting lever curve between the angles of heel of 30° and 45°	0.034 m rad
3	Value of maximum righting lever	G_FM
4	Angle of heel for maximum righting lever	60°
5	Initial metacentric height corrected for free surface (G_FM)	0.2 m

Table 2.5 Stability criteria to take into account the effects of wind (from Bureau Veritas 2016)

	Description	Value
1	Heeling arm at angle of static equilibrium	$\leq 60\%$ of GZ_{Max}
2	Area A1	$\geq 1.4 \times$ Area A2
3	Static equilibrium angle	$\leq 15°$

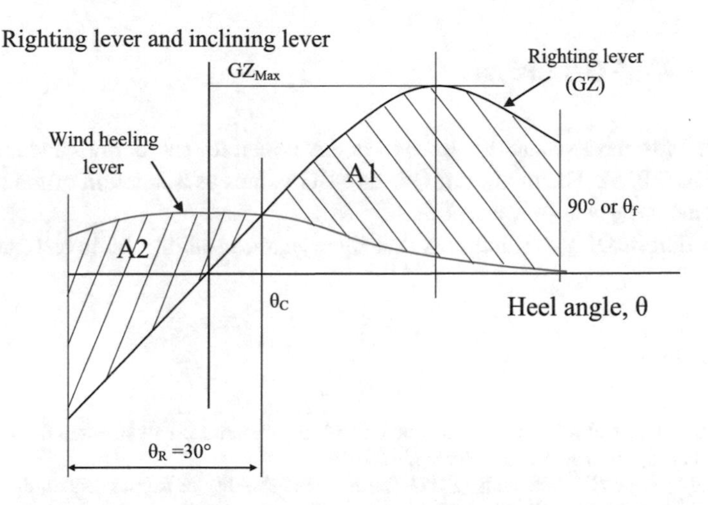

Fig. 2.18 Righting and wind heeling lever curves (from Bureau Veritas 2016)

downflooding, and area A2 is the area between the wind heeling lever curve and the righting lever curve from the angle of static equilibrium to an angle 30° less than this as shown in Fig. 2.18.

The wind heeling lever is calculated using a wind speed of 100 knots for submarines employed in very heavy conditions, and 80 knots for submarines not employed in storms, but having to be able to face other heavy weather conditions.

The wind heeling lever is calculated using Eq. 2.6.

$$\text{Wind heeling lever} = \frac{0.0195 V_{wind}^2 A_{wind} h \cos^2\theta}{1000\Delta} \tag{2.6}$$

In Eq. 2.6 V_{wind} is the wind speed in knots, A_{wind} is the area of windage above the waterline in m^2 and h is the vertical distance, in m, between the windage surface centre and the driftage surface centre (which may be considered at mid-draught).

In addition, Bureau Veritas (2016) gives guidance on accretion due to ice, and on criteria for damage stability, however these are beyond the scope of this book.

Table 2.6 Stability criteria (from DNV-GL 2015)

Displacement (t)	Surfaced	Submerged
	G_FM (m)	BG_F (m)
200–500	0.15	0.22
500–1000	0.18	0.27
1000–2000	0.20	0.32
>2000	0.22	0.35

2.8.3 DNV-GL Criteria

Criteria for transverse stability have been developed for submarines and are given in DNV-GL (2015). The minimum GM and BG values as a function of the size of the submarine are given in Table 2.6.

In addition, DNV-GL requires that the dynamic stability be investigated.

References

Veritas B (2016) Rules for the classification of naval submarines, Part B, Main design principles and stability. Bureau Veritas, September 2016

Crossland P, Pope CK, Machin S (2017) The dynamic stability of submarines on the surface. In: Proceedings of warship 2017 naval submarines and UUVs, Royal Institution of Naval Architects, Bath, UK

DNV-GL (2015) Chapter 1: Rules for classification: naval vessels—DNVGL-RU-NAVAL-Part 4, December 2015

IMO (2008) Resolution MSC.267(85) international code on intact stability (adopted on 4 December 2008)

Renilson M (2016) Hydrostatics. In: Compendium of ship hydrodynamics, practical tools and applications, Les Presses de l'ENSTA, January, 2016. ISBN-10: 2722509490, ISBN-13: 978-2722509498

Rawson KJ, Tupper EC (2001) Basic ship theory, 5th edn. Butterworth-Heinemann

Chapter 3
Manoeuvring and Control

Abstract The equations of motion for submarine manoeuvring are presented and discussed together with a non-linear coefficient based approach for determining the forces and moments on the submarine. Means of determining the coefficients using model tests, including a rotating arm and a planar motion mechanism, are detailed. In addition, the use of Computational Fluid Dynamics; and empirical techniques for determining the manoeuvring coefficients are discussed. Empirical equations for determining the manoeuvring coefficients are presented, and the results compared to published results from experiments. Issues associated with manoeuvring in the horizontal and vertical planes are explained, including: stability in the horizontal plane; the Pivot Point; heel during a turn, including snap roll; the effect of the sail, including the stern dipping effect; the Centre of Lateral Resistance; stability in the vertical plane; the Neutral Point; and the Critical Point, including the effect of speed, and issues at very low speed. Manoeuvring close to the surface, including surface suction, is discussed. Suggested criteria for stability in the horizontal and vertical planes, along with rudder and plane effectiveness are given. The concept of Safe Operating Envelopes, including Manoeuvring Limitation Diagrams and Safe Manoeuvring Envelopes together with the associated Standard Operating Procedures in event of credible failures are presented. Free running model experiments and manoeuvring trials, including submarine definitive manoeuvres and submarine trials procedures are discussed.

3.1 Introduction

The basic concepts behind the manoeuvring of a submarine are very similar to that of a surface ship. The main differences between a study of submarine manoeuvring and that of surface ship manoeuvring are that a submarine can manoeuvre in all six degrees of freedom, but is very unlikely to be required to manoeuvre whilst going astern.

The original version of this chapter was revised: Belated corrections have been incorporated. The erratum to this chapter is available at https://doi.org/10.1007/978-3-319-79057-2_8

M. Renilson, *Submarine Hydrodynamics*,
https://doi.org/10.1007/978-3-319-79057-2_3

As with surface ships, there are four different possible levels of motion stability:

(a) unstable;
(b) straight line stability (after a disturbance the boat remains on a straight line, but at a different heading from the initial heading);
(c) directional stability (after a disturbance the boat remains on the original heading, but is displaced from the initial path); and
(d) positional motion stability (after a disturbance the boat remains on the original path).

These are illustrated in Fig. 3.1. Note that directional stability can either include oscillations, prior to settling on a straight line, or not, as shown in Fig. 3.1c. The latter is referred to as critically damped, and is shown by the solid line.

An important aspect of the manoeuvring of a submarine is that the degree of manoeuvrability and motion stability required in the vertical plane may be different to that required in the horizontal plane. A normal military submarine has only a very limited range of operation in the vertical plane—typically only a few boat lengths. Above this it will break through the surface (broaching), and below that it will exceed the Deep Diving Depth (DDD), or hit the seabed. Thus, particularly for

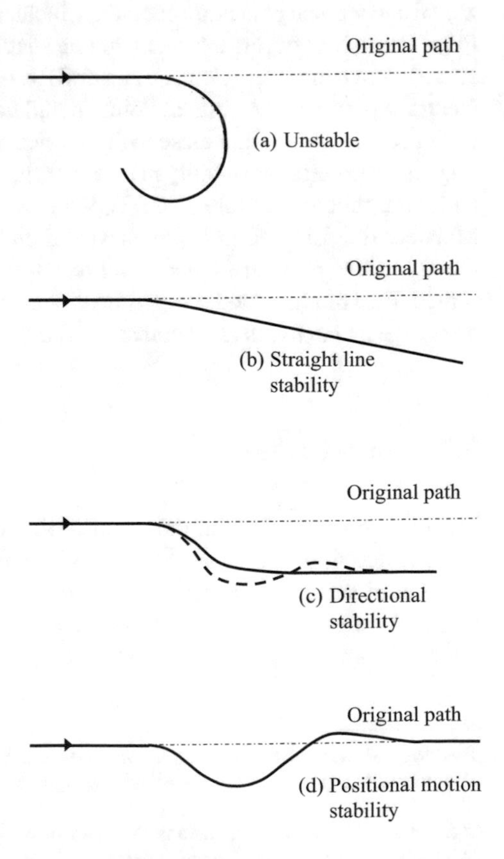

Fig. 3.1 Motion stability modes

high speed nuclear powered submarines, it is desirable to have a good degree of motion stability in the vertical plane. This may be of less importance for lower speed conventional boats where the ability to have a high degree of manoeuvrability in the vertical plane may give a tactical advantage when operating over an undulating seabed.

Hence, an important aspect for the submarine designer at an early stage in the design is to determine the level of manoeuvrability and motion stability required in each plane. Recommended values are given in Sect. 3.9.

Another important point is that with the controls fixed the degree of motion stability possible in the vertical plane is different to that in the horizontal plane. In the horizontal plane, the greatest possible level of motion stability with the controls fixed is straight line stability. With this level of stability, after being disturbed by a small deflection a submarine will return to a straight line motion, but not in the same direction as prior to the disturbance, as shown in Fig. 3.1b. To achieve the same direction it is necessary to have operating controls.

On the other hand, it is possible for a submarine to have directional stability in the vertical plane. With this level of stability, after being disturbed by a small deflection a submarine will return to the same direction. This is shown in Fig. 3.1c. This is possible because of the influence of the hydrostatic force, discussed in Chap. 2, which provides a pitch restoring moment.

3.2 Equations of Motion

The equations of motion for a submarine are similar to those for a surface ship, however they include all six degrees of freedom. For a submarine it is normal to take the origin as the longitudinal centre of gravity (LCG), rather than midships, as this simplifies the equations, and for a submarine this position is fixed (unlike for a surface ship). The axis system used is shown in the notation. Note that the origin is on the centreline, which is where the transverse centre of gravity is assumed to be. Positive directions are along the positive axes, and positive rotations are clockwise as seen from the origin looking along the positive direction of the axes.

The notation is given in Table 3.1, and in the notation section.

Table 3.1 Notation

	Position	Velocity	Force/moment
Surge	x	u	X
Sway	y	v	Y
Heave	z	w	Z
Roll	ϕ	p	K
Pitch	θ	q	M
Yaw	ψ	r	N
Appendage	δ		
Propulsor		n	

The equations of motion are based on Newton's Second Law: Force = Mass × Acceleration

In this case the force, the left hand side of the equation, is the hydrodynamic force acting on the submarine, and the right hand side is the rigid body dynamics. This equation is transformed into body fixed axes, and the right hand side of the equation is given as follows:

$$X = \mathrm{m}\left[\dot{u} - vr + wq - x_G\left(q^2 + r^2\right) + y_G(pq - \dot{r}) + z_G(pr + \dot{q})\right] \tag{3.1}$$

$$Y = \mathrm{m}\left[\dot{v} - wp + ur + x_G(qp + \dot{r}) - y_G\left(r^2 + p^2\right) + z_G(qr - \dot{p})\right] \tag{3.2}$$

$$Z = \mathrm{m}\left[\dot{w} - uq + vp + x_G(rp - \dot{q}) + y_G(rq + \dot{p}) - z_G\left(p^2 + q^2\right)\right] \tag{3.3}$$

$$\begin{aligned} K = {} & I_{xx}\dot{p} + \left(I_{zz} - I_{yy}\right)qr - (\dot{r} + pq)I_{zx} + \left(r^2 - q^2\right)I_{yz} + (pr - \dot{q})I_{xy} \\ & + \mathrm{m}[y_G(\dot{w} - uq + vp) - z_G(\dot{v} - wp + ur)] \end{aligned} \tag{3.4}$$

$$\begin{aligned} M = {} & I_{yy}\dot{q} + (I_{xx} - I_{zz})rp - (\dot{p} + qr)I_{xy} + \left(p^2 - r^2\right)I_{zx} + (qp - \dot{r})I_{yz} \\ & - \mathrm{m}[x_G(\dot{w} - uq + vp) - z_G(\dot{u} - vr + wq)] \end{aligned} \tag{3.5}$$

$$\begin{aligned} N = {} & I_{zz}\dot{r} + \left(I_{yy} - I_{xx}\right)pq - (\dot{q} + rp)I_{yz} + \left(q^2 - p^2\right)I_{xy} + (rq - \dot{p})I_{zx} \\ & + \mathrm{m}[x_G(\dot{v} - wp + ur) - y_G(\dot{u} - vr + wq)] \end{aligned} \tag{3.6}$$

If the origin of the axes is taken at the position of the longitudinal, and transverse centre of gravity, then both x_G and y_G will be equal to zero, simplifying these equations.

X, Y, Z, K, M, and *N* are the total hydrodynamic surge, sway, and heave forces, and roll, pitch and yaw moments respectively. If these hydrodynamic forces and moments can be determined as functions of time for a manoeuvring submarine, then the manoeuvre can be simulated. In addition, if the effects of geometry on these forces and moments are understood then this can be used to assist in the design of the submarine.

3.3 Hydrodynamic Forces—Steady State Assumption

3.3.1 Coefficient Based Model

One approach to determining the hydrodynamic forces and moments on a manoeuvring submarine is to assume that at any point in time these forces and moments are functions of the motions (velocities and accelerations), propeller rpm, and appendage angles, at that point in time. This is a similar approach to that used for surface ships.

As with surface ships, the relationship between each motion variable and the resultant force or moment can be represented by a mathematical model comprising a series of coefficients. The resulting forces and moments due to each of these are then added to give the total force or moment on the submarine at that point in time. The choice of which coefficients, and hence which mathematical model, to use will depend on experience. It is normal for a single mathematical model to be used by a given organisation to represent different submarines. Once the mathematical model representing the forces and moments has been selected, different submarines, or changes to the shape of a given submarine, can be represented by changing the values of the individual coefficients.

It is important to recognise that as different organisations may use different mathematical models it is not necessarily possible to compare the values of coefficients between different organisations. Also, as improvements in understanding are achieved, and the mathematical model updated, care needs to be taken to ensure that legacy coefficient sets are retained for past submarines.

A typical mathematical model to represent the three forces and three moments as functions of the current motion of the submarine is given in Eqs. 3.7–3.12, Gertler and Hagen (1967). These equations were revised by Feldman (1979), however the original Gertler and Hagen equations are commonly used in the submarine community.

$$\begin{aligned}
X = {} & \frac{1}{2}\rho L^4\left[X'_{qq}q^2 + X'_{rr}r^2 + X'_{rp}rp\right] \\
& + \frac{1}{2}\rho L^3\left[X'_{\dot{u}}\dot{u} + X'_{vr}vr + X'_{wq}wq\right] \\
& + \frac{1}{2}\rho L^2\left[X'_{uu}u^2 + X'_{vv}v^2 + X'_{ww}w^2 + X'_{\delta R\delta R}u^2\delta_R^2 + X'_{\delta s\delta s}u^2\delta_S^2 + X'_{\delta B\delta B}u^2\delta_B^2\right] \\
& + \frac{1}{2}\rho L^2\left[a_i u^2 + b_i u u_c + c_i u_c^2\right] \\
& - (\mathrm{W} - \mathrm{B})\sin\theta \\
& + \frac{1}{2}\rho L^2\left[X'_{vv\eta}v^2 + X'_{ww\eta}w^2 + X'_{\delta R\delta R\eta}\delta_R^2 u^2 + X'_{\delta s\delta s\eta}\delta_s^2 u^2\right](\eta - 1)
\end{aligned} \tag{3.7}$$

$$\begin{aligned}
Y = {} & \frac{1}{2}\rho L^4\left[Y'_{\dot{r}}\dot{r} + Y'_{\dot{p}}\dot{p} + Y'_{p|p|}p|p| + Y'_{pq}pq + Y'_{qr}qr\right] \\
& + \frac{1}{2}\rho L^3\left[Y'_{\dot{v}}\dot{v} + Y'_{vq}vq + Y'_{wp}wp + Y'_{wr}wr\right] \\
& + \frac{1}{2}\rho L^3\left[Y'_r ur + Y'_p up + Y'_{|r|\delta R}u|r|\delta_R + Y'_{v|r|}\frac{v}{|v|}\left|(v^2 + w^2)^{\frac{1}{2}}\right||r|\right] \\
& + \frac{1}{2}\rho L^2\left[Y'_* u^2 + Y'_v uv + Y'_{v|v|}v\left|(v^2 + w^2)^{\frac{1}{2}}\right| + Y'_{vw}vw + Y'_{\delta R}u^2\delta_R\right] \\
& + (\mathrm{W} - \mathrm{B})\cos\theta\sin\phi \\
& + \frac{1}{2}\rho L^3 Y'_{r\eta}ur(\eta - 1) \\
& + \frac{1}{2}\rho L^2\left[Y'_{v\eta}uv + Y'_{v|v|\eta}v\left|(v^2 + w^2)^{\frac{1}{2}}\right| + Y'_{\delta R\eta}u^2\delta_R\right](\eta - 1)
\end{aligned} \tag{3.8}$$

$$
\begin{aligned}
Z = &\frac{1}{2}\rho L^4\left[Z'_{\dot{q}}\dot{q} + Z'_{pp}p^2 + Z'_{rr}r^2 + Z'_{rp}rp\right] \\
&+ \frac{1}{2}\rho L^3\left[Z'_{\dot{w}}\dot{w} + Z'_{vr}vr + Z'_{vp}vp + Z'_{q}uq + Z'_{|q|\delta s}u|q|\delta_s\right] \\
&+ \frac{1}{2}\rho L^3\left[Z'_{w|q|}\frac{w}{|w|}\left|\left(v^2+w^2\right)^{\frac{1}{2}}\right||q|\right] \\
&+ \frac{1}{2}\rho L^2\left[Z'_{*}u^2 + Z'_{v}uv + Z'_{w}uw + Z'_{w|w|}w\left|\left(v^2+w^2\right)^{\frac{1}{2}}\right|\right] \\
&+ \frac{1}{2}\rho L^2\left[Z'_{|w|}u|w| + Z'_{ww}\left|w\left(v^2+w^2\right)^{\frac{1}{2}}\right|\right] \\
&+ \frac{1}{2}\rho L^2\left[Z'_{vv}v^2 + Z'_{\delta s}u^2\delta_s + Z'_{\delta B}u^2\delta_B\right] \\
&+ (\mathrm{W}-\mathrm{B})\cos\theta\cos\phi \\
&+ \frac{1}{2}\rho L^3 Z'_{q\eta}uq(\eta-1) \\
&+ \frac{1}{2}\rho L^2\left[Z'_{w\eta}uw + Z'_{w|w|\eta}w\left|\left(v^2+w^2\right)^{\frac{1}{2}}\right| + Z'_{\delta s\eta}\delta_s u^2\right](\eta-1)
\end{aligned}
\tag{3.9}
$$

$$
\begin{aligned}
K = &\frac{1}{2}\rho L^5\left[K'_{\dot{p}}\dot{p} + K'_{\dot{r}}\dot{r} + K'_{qr}qr + K'_{pq}pq + K'_{p|p|}p|p|\right] \\
&+ \frac{1}{2}\rho L^4\left[K'_{p}up + K'_{r}ur + K'_{\dot{v}}\dot{v} + K'_{vq}vq + K'_{wp}wp + K'_{wr}wr\right] \\
&+ \frac{1}{2}\rho L^3\left[K'_{*}u^2 + K'_{v}uv + K'_{v|v|}v\left|\left(v^2+w^2\right)^{\frac{1}{2}}\right| + K'_{vw}vw + K'_{\delta R}u^2\delta_R\right] \\
&+ (y_G\mathrm{W} - y_B\mathrm{B})\cos\theta\cos\phi - (z_G\mathrm{W} - z_B\mathrm{B})\cos\theta\sin\phi \\
&+ \frac{1}{2}\rho L^3 K'_{*\eta}u^2(\eta-1)
\end{aligned}
\tag{3.10}
$$

$$
\begin{aligned}
M = &\frac{1}{2}\rho L^5\left[M'_{\dot{q}}\dot{q} + M'_{pp}p^2 + M'_{rr}r^2 + M'_{rp}rp + M'_{q|q|}q|q|\right] \\
&+ \frac{1}{2}\rho L^4\left[M'_{\dot{w}}\dot{w} + M'_{vr}vr + M'_{vp}vp\right] \\
&+ \frac{1}{2}\rho L^4\left[M'_{q}uq + M'_{|q|\delta s}u|q|\delta_s + M'_{|w|q}\left|\left(v^2+w^2\right)^{\frac{1}{2}}\right|q\right] \\
&+ \frac{1}{2}\rho L^3\left[M'_{*}u^2 + M'_{w}uw + M'_{w|w|}w\left|\left(v^2+w^2\right)^{\frac{1}{2}}\right|\right] \\
&+ \frac{1}{2}\rho L^3\left[M'_{|w|}u|w| + M'_{ww}\left|w\left(v^2+w^2\right)^{\frac{1}{2}}\right|\right] \\
&+ \frac{1}{2}\rho L^3\left[M'_{vv}v^2 + M'_{\delta s}u^2\delta_s + M'_{\delta B}u^2\delta_B\right] \\
&- (x_G\mathrm{W} - x_B\mathrm{B})\cos\theta\cos\phi - (z_G\mathrm{W} - z_B\mathrm{B})\sin\theta \\
&+ \frac{1}{2}\rho L^4 M'_{q\eta}uq(\eta-1) \\
&+ \frac{1}{2}\rho L^3\left[M'_{w\eta}uw + M'_{w|w|\eta}w\left|\left(v^2+w^2\right)^{\frac{1}{2}}\right| + M'_{\delta s\eta}u^2\delta_s](\eta-1)\right]
\end{aligned}
\tag{3.11}
$$

$$
\begin{aligned}
N = & \frac{1}{2}\rho L^5\left[N'_{\dot{r}}\dot{r} + N'_{\dot{p}}\dot{p} + N'_{pq}pq + N'_{qr}qr + N'_{r|r|}r|r|\right] \\
& + \frac{1}{2}\rho L^4\left[N'_{\dot{v}}\dot{v} + N'_{wr}wr + N'_{wp}wp + N'_{vq}vq\right] \\
& + \frac{1}{2}\rho L^4\left[N'_{p}up + N'_{r}ur + N'_{|r|\delta R}u|r|\delta_R + N'_{|v|r}\left|\left(v^2 + w^2\right)^{\frac{1}{2}}\right|r\right] \\
& + \frac{1}{2}\rho L^3\left[N'_{*}u^2 + N'_{v}uv + N'_{v|v|}v\left|\left(v^2 + w^2\right)^{\frac{1}{2}}\right| + N'_{vw}vw + N'_{\delta R}u^2\delta_R\right] \\
& + (x_G\mathrm{W} - x_B\mathrm{B})\cos\theta\sin\phi + (y_G\mathrm{W} - y_B\mathrm{B})\sin\theta \\
& + \frac{1}{2}\rho L^4 N'_{r\eta}ur(\eta - 1) \\
& + \frac{1}{2}\rho L^3\left[N'_{v\eta}uv + N'_{v|v|\eta}v\left|\left(v^2 + w^2\right)^{\frac{1}{2}}\right| + N'_{\delta R\eta}u^2\delta_R\right](\eta - 1)
\end{aligned}
\tag{3.12}
$$

When the submarine is travelling at its self-propulsion speed, η will be equal to 1, and hence the last terms in each of these equations will be zero. For manoeuvres close to steady state this is usually an accepted approximation.

Note that these equations are relevant to a submarine with a cruciform stern control configuration. See Sect. 6.4.3 for the changes that are required for a submarine with an X-form configuration.

If the origin is at the longitudinal position of the centre of gravity (hence $x_G = 0$) the equations will be significantly simplified. However, the location of the origin will affect the values of some of the coefficients. Their values can be transformed using Eqs. 3.13–3.18.

$$Y_{r_G} = Y_{r_O} - Y_v x_G \tag{3.13}$$

$$N_{r_G} = N_{r_O} + Y_v x_G^2 - N_{v_O} x_G - Y_{r_O} x_G \tag{3.14}$$

$$N_{v_G} = N_{v_O} - Y_v x_G \tag{3.15}$$

$$Z_{q_G} = Z_{q_O} + Z_w x_G \tag{3.16}$$

$$M_{q_G} = M_{q_O} + Z_w x_G^2 + M_{w_O} x_G + Z_{q_O} x_G \tag{3.17}$$

$$M_{w_G} = M_{w_O} + Z_w x_G \tag{3.18}$$

where the subscript "G" refers to the origin at the longitudinal centre of gravity, and subscript "O" refers to the origin at a distance x_G from the longitudinal centre of gravity. The value of Y_v is independent on the longitudinal position of the origin.

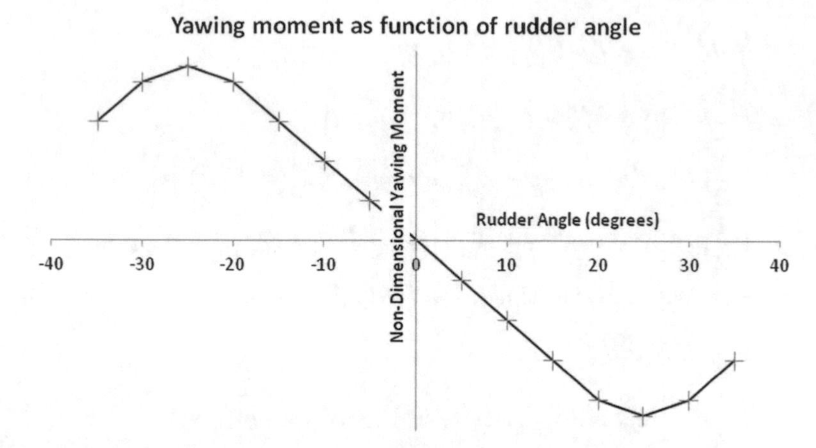

Fig. 3.2 Yawing moment as a function of rudder angle

3.3.2 *Look-up Tables*

An alternative approach to quantifying the relationship between each motion variable and the resulting hydrodynamic force or moment is to use a series of look-up tables, Jensen et al. (1993).

In principle this makes it easier to 'fit' measured data as there is complete freedom as to the form of the function, and it is not necessary to force the data to fit a particular representation using a predetermined expression. This approach can be particularly useful for some relationships, where the function of the force or moment, in terms of the motion parameter, is not a clearly determined smooth curve.

An example where this may be appropriate is the relationship between an appendage angle and the resulting lift force. The look-up table approach makes it much easier to represent 'stall' of the appendage than a coefficient approach, as shown in Fig. 3.2. The data points are shown as crosses in this figure, and the model uses straight line interpolation to determine the non-dimensional yawing moment at any angle.

A hybrid approach may be appropriate. This makes use of a number of coefficients to represent the functions between most of the forces or moments and the relative motion parameters, along with a few look-up tables for the functions for other motion parameters.

3.3.3 *Sensitivity of Individual Coefficients*

Not all the coefficients in Eqs. 3.7–3.12 are of equal importance. Depending on the particular manoeuvre, the various coefficients have different levels of significance. It is important to understand the sensitivity of the individual coefficients to ensure that they are predicted to the necessary level of accuracy. Equally, knowing that a particular coefficient will have negligible influence on the manoeuvres of interest means that less effort can be exerted on its prediction.

Sen (2000) conducted a useful investigation into the sensitivity of the various coefficients for submerged bodies. He defined the sensitivity of a particular coefficient using Eq. 3.19.

$$S = \frac{\Delta R/R^{*}}{\Delta H/H^{*}} \tag{3.19}$$

In Eq. 3.19 S is the sensitivity index of a particular coefficient, ΔR is the difference in path, R^{*}, and ΔH is the change in the coefficient with respect to the base value, H^{*}. The higher the value of S the more sensitive the manoeuvre is to the particular coefficient.

Sen (2000) conducted a range of simulations for the following three manoeuvres:

i. Overshoot manoeuvre in the vertical plane;
ii. Overshoot manoeuvre in the horizontal plane; and
iii. Turning circle manoeuvre in the horizontal plane.

These manoeuvres are described in Sect. 3.12.2. Note the similarity between the zig-zag manoeuvre and the overshoot manoeuvre, where the overshoot manoeuvre is the first cycle in the zig-zag.

Sen (2000) conducted the simulations using a range of speeds and stern plane angles for two different underwater bodies:

i. A typical submarine configuration; and
ii. An axisymmetric slender body.

The Sen Sensitivity Index for the 10 most important coefficients is given in Table 3.2 for the three different manoeuvres for the submarine.

Table 3.2 Sen sensitivity index for submarine (taken from Sen 2000)

Overshoot in vertical plane		Overshoot in horizontal plane		Turning circle	
Coeff.	S	Coeff.	S	Coeff.	S
M_q	4.914	N_r	3.958	$N_{\delta R}$	1.527
$M_{\delta S}$	3.154	$Y_{\delta R}$	1.607	N_v	1.125
$Z_{\delta S}$	2.305	$Y_{\dot{v}}$	1.479	N_r	1.018
Z_q	2.025	Y_r	1.430	Y_v	0.468
$Z_{\dot{w}}$	1.597	Y_v	1.406	Y_r	0.462
Z_w	1.290	N_v	1.292	$N_{\dot{r}}$	0.349
Z_*	1.203	$N_{\delta R}$	1.225	X_{vr}	0.311
$Z_{\lvert w \rvert}$	1.046	K_{vr}	1.223	$Y_{\delta R}$	0.292
$Z_{\dot{q}}$	0.988	Y_p	1.216	K_{vr}	0.279
M_w	0.979	$Y_{\dot{p}}$	1.214	$M_{\delta S}$	0.260

The values given in Table 3.2 are indicative only. The relative importance of the different coefficients is likely to change depending on the geometry. For example, Sen found that the coefficients for the inertia forces and moments were more important for the axisymmetric body than for the submarine.

However, Table 3.2 does give a useful guide as to the most important coefficients for predicting each of the three manoeuvres. Note that the nonlinear coefficients have very little influence on the main measured quantities from these manoeuvres, however they may well affect the detailed trajectory, particularly for more extreme manoeuvres such as those required to assess the Safe Operating Envelope—see Sect. 3.10. In particular, the small vertical forces which arise due to the asymmetry when turning (see Sect. 3.6) may need to be modelled correctly. Although these may not have a significant effect on the turning circle directly, they can have an influence on the heave/pitch behaviour, and hence on the required stern plane angle when turning.

3.4 Determination of Coefficients

3.4.1 Model Tests

3.4.1.1 General

The most common way of determining the values of the coefficients required for the approach discussed in Sect. 3.3 is to conduct captive model tests. The approach is very similar to that used for surface ship models.

Normally fairly large models are used (5–6 m long) as even at such a large scale the appendages are actually quite small, with low local Reynolds numbers. In addition, as scale effects on the shedding of vortices are not fully understood, the generally accepted procedure is to use as large a model as possible and to neglect scale effects. Turbulence stimulation is normally fitted to the hull and appendages.

As a deeply submerged submarine does not interact with the surface it is not necessary to conduct captive model experiments at the correct Froude number. Thus, it is only Reynolds number that is of importance.

In principle, tests can be conducted in either water, or air, in a towing tank or a water/wind tunnel. A common procedure is to test in a large towing tank, with the model supported inverted from the carriage using struts as shown in Fig. 3.3.

When testing in a towing tank it is important to recognise the presence of the water surface. This means that the speed needs to be limited to prevent waves occurring, with the resulting Froude number effects. Most facilities have a common speed that they always test their submarine models, to give consistency. For example, in the QinetiQ facility at Haslar, UK, the normal test speed is 10 ft per second, which has been used for historical reasons. This, together with a standard

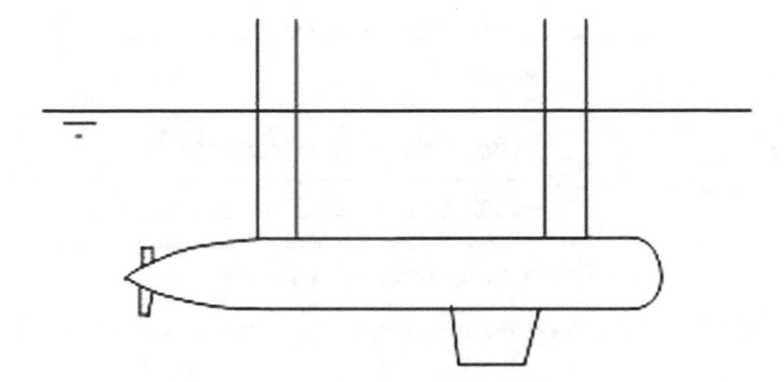

Fig. 3.3 Typical set up for captive model tests in a towing tank

turbulence stimulation method, and a similar sized model, means that any scale effects etc. will be consistent for all tests—an important aspect of tank testing.

The effect of the support struts needs to be considered. Although the hydrodynamic forces are measured inside the model, and hence the forces on the struts are not included in the measurements, the presence of the struts can influence the flow around the model. For this reason the model is tested either inverted, or on its side, depending on which coefficients are being investigated.

It is possible to use a sting type mount, as shown in Fig. 3.4, however this generally means that the propulsor cannot be included. As the propulsor has a significant influence on the flow over the stern of the submarine (see Chap. 4) care needs to be taken with this approach.

The approach used for captive model testing is to confine the model to a given motion, and then to measure the resulting forces.

3.4.1.2 Tests in Translation (Sway/Heave)

To obtain the values of the coefficients which represent the forces and moments as functions of sway velocity, such as Y_v, N_v, etc., the model is tested on its side with the sway velocity being generated by adjusting the angle of the model in the vertical plane. This avoids the need for the struts to be at an angle to the flow, as would be required if the model were rotated in the horizontal plane. This minimises the hydrodynamic disturbance that they create. However, as most towing tanks are wider than they are deep this does have the disadvantage of increasing the effective blockage, compared to adjusting the angle in the horizontal plane. Note that this technique will not work if the effect of the presence of the water surface on manoeuvring in the horizontal plane is being investigated—see Sect. 3.8. For this case it is necessary to use a sting type mount (Fig. 3.4) and adjust the angle of the model in the horizontal plane.

To obtain the values of the coefficients which represent the forces and moments as functions of heave velocity, such as Z_w, M_w, etc., the model is tested inverted, as

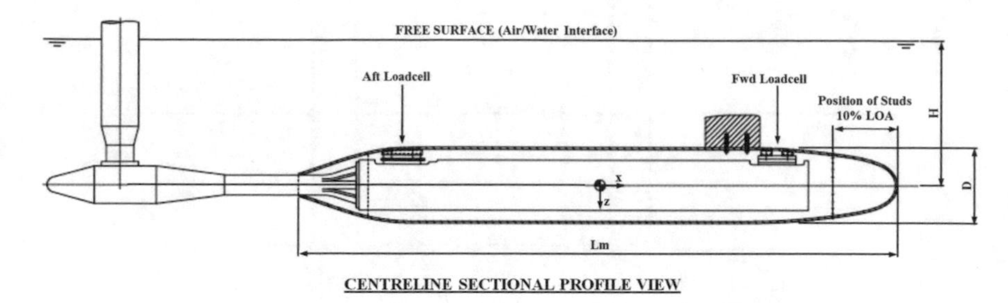

Fig. 3.4 Set up for captive model tests using a sting support (taken from Renilson et al. 2011)

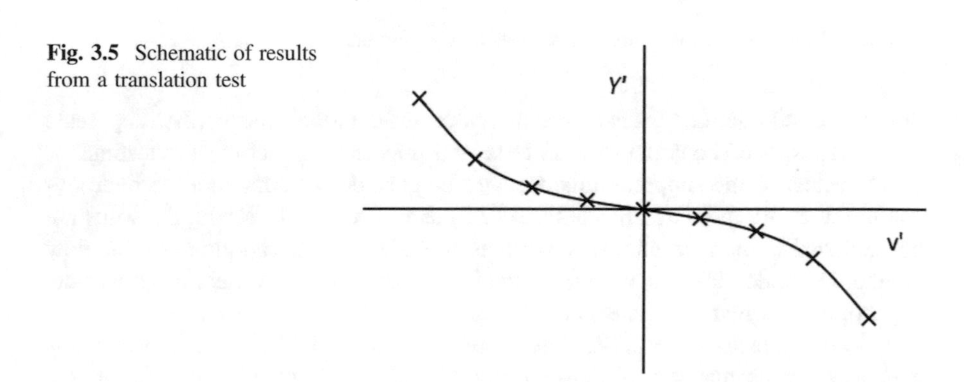

Fig. 3.5 Schematic of results from a translation test

shown in Fig. 3.3, with the heave velocity being generated by adjusting the angle of the model in the vertical plane. Again, this avoids the need for the struts to be at an angle to the flow.

A schematic of the typical results from such an experiment, where the non-dimensional side force (Y') is plotted as a function of the non-dimensional sway velocity (v') is given in Fig. 3.5.

In this case, with the propeller revolutions set to the self-propulsion speed, the hydrodynamic side force is represented by Eq. 3.20, which is simplified from Eq. 3.8.

$$Y = \frac{1}{2}\rho L^2 \left[Y'_* u^2 + Y'_v uv + Y'_{v|v|} v|v| \right] \tag{3.20}$$

In Eq. 3.20 there are three unknown terms which can be obtained from this experiment: Y'_*, Y'_v, and $Y'_{v|v|}$. Y'_* is due to an asymmetry—the results not passing through $Y' = 0$ at $v' = 0$. The remaining two coefficients are obtained from a "fit" to the data, with Y'_v representing the linear characteristic of the data (dominant at low values of v') and $Y'_{v|v|}$ representing the non-linear characteristic. Note that the results in Fig. 3.5 are skew symmetric, hence the need for an "odd" term, such as the $v|v|$ term, rather than v^2. An alternative would be to use v^3, which also provides skew symmetry. However, as hydrodynamic forces tend to be proportional to velocity squared, the $v|v|$ term is often preferred, as in the original work of Gertler and Hagen (1967).

The values of the coefficients which represent the forces and moments as functions of appendage angles can be obtained by setting the relevant appendages to the required angles. However, Reynolds number effects can influence the lift and drag on the appendages, as the model scale Reynolds number will be much lower than the full scale value. Hence care has to be taken when interpreting the results. In addition, the appendages operate in the model hull's boundary layer, which is much larger at model scale than full scale, due to Reynolds number effects, and this may also influence the results.

The cross coupling effects between yaw or heave velocity and appendage angles can also be obtained, as can the effect of propulsor rpm on the forces and moments due to the appendage angles. Again, care needs to be taken with scale effects due to testing at the wrong Reynolds number.

3.4.1.3 Rotating Arm

As with surface ship models, to obtain the values of the coefficients which represent the forces and moments as functions of yaw velocity, it is necessary to test the models in rotation, using a rotating arm. This is done in the horizontal plane, and the model is tested inverted, as shown in Fig. 3.6.

The coefficients which represent the forces and moments as functions of the pitch velocity can be obtained with the model tested on its side.

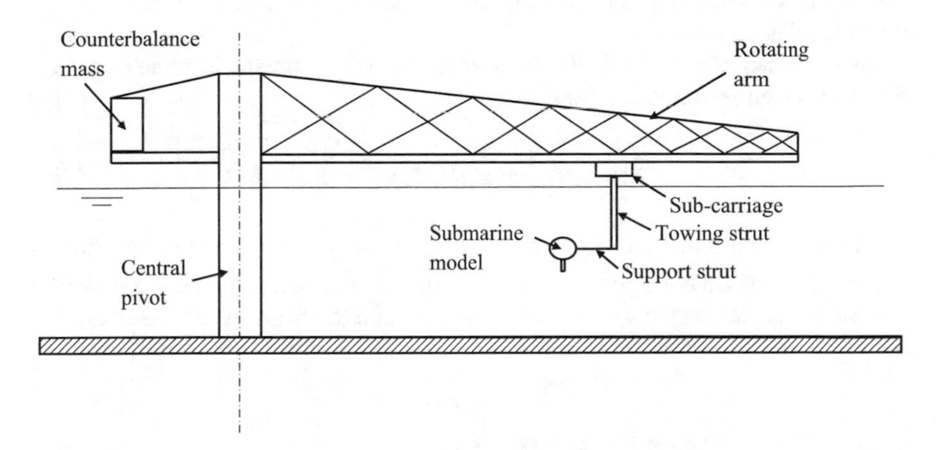

Fig. 3.6 Typical set up for captive model tests using a rotating arm

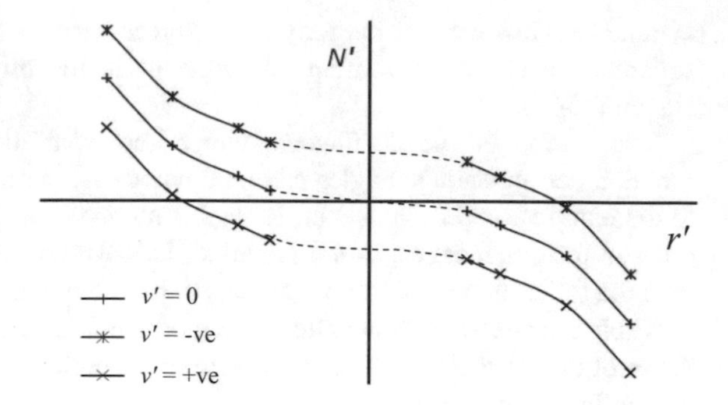

Fig. 3.7 Schematic of results from rotating arm

The arm is fitted with a sub-carriage, as shown in Fig. 3.6, which makes it possible to test at different radii, and hence different values of q (model on side) or r (model inverted).

Of course, as with surface ship captive model tests, the rotating arm can be used to obtain the values of the coefficients which represent the forces and moments as functions of sway and heave velocity by carrying out cross plots. It can also be used to obtain the values of cross coefficients. A sketch of the results from a rotating arm is given in Fig. 3.7. In this case the model is inverted, and being tested at a range of values of r'. The non-dimensional moment N' has been plotted as a function of r' for a range of values of v'.

However, a major difficulty with the rotating arm is that it is not possible to test at small values of q', or r', which would require very large radii (q', $r' = 0$ is a straight line). This is shown in Fig. 3.7. In the figure the experimentally obtained points are joined by a dotted line across $r' = 0$ where experiments are not possible. The linear coefficient is the gradient of the line at $r' = 0$ and the difficulty in obtaining this can be seen.

In this case, when $v' = 0$ the hydrodynamic yaw moment is represented by Eq. 3.21 which is simplified from Eq. 3.12.

$$N = \frac{1}{2}\rho L^5\left[N'_{r|r|}r|r|\right] + \frac{1}{2}\rho L^4\left[N'_r ur\right] + \frac{1}{2}\rho L^3\left[N'_* u^2\right] \tag{3.21}$$

The unknown terms are: $N'_{r|r|}$, N'_r, and N'_*. These can be obtained by fitting a curve to the experimental points in Fig. 3.7. However, as noted above, the difficulty in defining the curve at low values of r' makes it hard to obtain an accurate value for N'_r.

3.4.1.4 Planar Motion Mechanism

As with surface ships, a Planar Motion Mechanism (PMM) can be used to obtain the added masses, and also to obtain the values of the coefficients which represent the forces and moments as functions of the rotary motions in a towing tank without

Fig. 3.8 Typical set up for captive model tests using a VPMM (courtesy QinetiQ Limited, © Copyright QinetiQ Limited 2017)

requiring a rotating arm. Generally for submarines this is done in the vertical plane, using a Vertical Planar Motion Mechanism (VPMM), as shown in Fig. 3.8.

The tests can be carried out in two different regimes:

(a) pure translation; and
(b) pure rotation.

When the model is inverted, pure translation gives pure heave, and pure rotation gives pure pitch. When the model is on its side, pure translation gives pure sway, and pure rotation gives pure yaw.

(a) Pure Heave

For the case with the model inverted, in pure heave, with the propulsor revolutions set at the self-propulsion point, the measured total force on the struts as a function of time, $Z_m(t)$, obtained by combining Eq. 3.3 (the rigid body component) with Eq. 3.9 (the hydrodynamic component) is given by Eq. 3.22.

$$\begin{aligned} Z_m(t) = & \frac{1}{2}\rho L^3\left[(Z'_{\dot{w}} - m')\dot{w}\right] \\ & + \frac{1}{2}\rho L^2\left[Z'_{*}u^2 + Z'_{w}uw + Z'_{w|w|}w|w|\right] \\ & + \frac{1}{2}\rho L^2\left[Z'_{|w|}u|w| + Z'_{ww}w^2\right] \end{aligned} \tag{3.22}$$

Linearizing, and ignoring the terms due to asymmetry, this simplifies to Eq. 3.23.

$$Z_m(t) = \frac{1}{2}\rho L^3\left[(Z'_{\dot{w}} - m')\dot{w}\right] + \frac{1}{2}\rho L^2\left[Z'_{w}uw\right] \tag{3.23}$$

In a similar way the linearized pitch moment on the struts for pure heave is given by Eq. 3.24.

$$M_m(t) = \frac{1}{2}\rho L^4\left[(M'_{\dot{w}} + m'x'_G)\dot{w}\right] + \frac{1}{2}\rho L^3\left[M'_{w}uw\right] \tag{3.24}$$

If the translation motion is sinusoidal, as given by Eq. 3.25 then the values of w and $\dot{w}$ are given by Eqs. 3.26 and 3.27 respectively.

$$z = z_0 \sin\omega t \tag{3.25}$$

$$w = z_0\omega \cos\omega t \tag{3.26}$$

$$\dot{w} = -z_0\omega^2 \sin\omega t \tag{3.27}$$

For motions which are sinusoidal, a linear system will give sinusoidal force output, with a phase shift. Thus, the measured force and moment can be represented by an in phase and an out of phase component, as given in Eqs. 3.28 and 3.29, where the dimensional values of the forces and moments are given.

$$Z_m(t) = Z_{in}\sin\omega t + Z_{out}\cos\omega t \tag{3.28}$$

$$M_m(t) = M_{in}\sin\omega t + M_{out}\cos\omega t \tag{3.29}$$

Thus, the non-dimensional linear coefficients can be obtained from Eqs. 3.30–3.33.

$$Z'_w = \frac{Z_{out}}{\frac{1}{2}\rho L^2 u z_o \omega} \tag{3.30}$$

$$(Z'_{\dot{w}} - m') = -\frac{Z_{in}}{\frac{1}{2}\rho L^3 z_o \omega^2} \tag{3.31}$$

$$M'_w = \frac{M_{out}}{\frac{1}{2}\rho L^3 u z_0 \omega} \tag{3.32}$$

$$(M'_{\dot{w}} - m'x'_G) = -\frac{M_{in}}{\frac{1}{2}\rho L^4 z_o \omega^2} \tag{3.33}$$

The coefficients determined in this way are frequency dependent, and so it may be necessary to extrapolate to zero frequency to obtain the steady state values which are used in Eqs. 3.7–3.12. For deeply submerged submarine models this is often not required, but should be considered if appropriate.

(b) Pure Pitch

For pure pitch, with the model inverted, the motion is given by Eqs. 3.34–3.36.

$$\theta = \theta_0 \sin\omega t \tag{3.34}$$

$$q = \theta_0 \omega \cos\omega t \tag{3.35}$$

$$\dot{q} = -\theta_0 \omega^2 \sin\omega t \tag{3.36}$$

The coefficients can be obtained from Eqs. 3.37–3.40.

$$(Z'_q + m') = \frac{Z_{out}}{\frac{1}{2}\rho L^3 u \theta_o \omega} \tag{3.37}$$

$$(Z'_{\dot{q}} + m'x'_G) = -\frac{Z_{in}}{\frac{1}{2}\rho L^4 \theta_o \omega^2} \tag{3.38}$$

$$(M'_q - m'x'_G) = \frac{M_{out}}{\frac{1}{2}\rho L^4 u \theta_0 \omega} \tag{3.39}$$

$$(M'_{\dot{q}} - I'_{yy}) = -\frac{M_{in}}{\frac{1}{2}\rho L^5 \theta_o \omega^2} \tag{3.40}$$

(c) Pure Sway

For pure sway, with the model on its side, the coefficients can be obtained from Eqs. 3.41–3.44.

$$Y'_v = \frac{Y_{out}}{\frac{1}{2}\rho L^2 u y_0 \omega} \tag{3.41}$$

$$(Y'_{\dot{v}} - m') = -\frac{Y_{in}}{\frac{1}{2}\rho L^3 y_o \omega^2} \tag{3.42}$$

$$N'_v = \frac{N_{out}}{\frac{1}{2}\rho L^3 u y_0 \omega} \tag{3.43}$$

$$(N'_{\dot{v}} - m'x'_G) = -\frac{N_{in}}{\frac{1}{2}\rho L^4 y_o \omega^2} \tag{3.44}$$

(d) Pure Yaw

For pure yaw, with the model on its side, the coefficients can be obtained from Eqs. 3.45–3.48.

$$(Y'_r - m') = \frac{Y_{out}}{\frac{1}{2}\rho L^3 u \psi_o \omega} \tag{3.45}$$

$$(Y'_{\dot{r}} - m'x'_G) = -\frac{Y_{in}}{\frac{1}{2}\rho L^4 \psi_o \omega^2} \tag{3.46}$$

$$(N'_r - m'x'_G) = \frac{N_{out}}{\frac{1}{2}\rho L^4 u \psi_0 \omega} \tag{3.47}$$

$$(N'_{\dot{r}} - I'_{zz}) = -\frac{N_{in}}{\frac{1}{2}\rho L^5 \psi_o \omega^2} \tag{3.48}$$

In the submarine design process the VPMM is usually used initially to determine the linear coefficients, which are those that affect the straight line stability of the submarine (in both the horizontal and vertical planes). This is used to make any refinements to the shape (if possible) and to determine the required size of the appendages. Once the design has moved to the next stage further tests are conducted in both the towing tank and the rotating arm facility.

A detailed description specifically for submarines using a VPMM is given in Booth and Bishop (1973).

3.4.1.5 Marine Dynamics Test Facility

An alternative approach to the PMM is to use a single mechanism to provide motion in all six degrees of freedom. Such a device was developed by the National Research Council, Canada, known as a Marine Dynamics Test Facility (MDTF) and

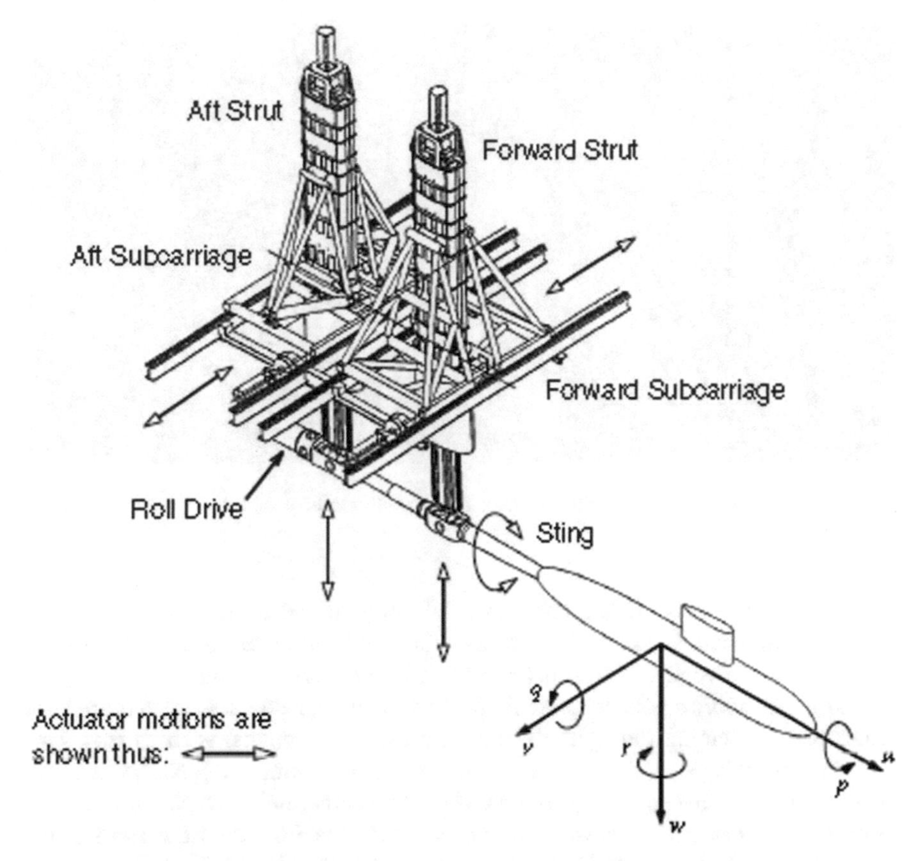

Fig. 3.9 Schematic of the marine dynamics test facility (reproduced with permission from the National Research Council of Canada)

is shown in Fig. 3.9, reproduced with permission from the National Research Council of Canada.

The control system for the MDTF enables it to perform all kinds of motions, including pure or combined manoeuvres, as discussed in Mackay et al. (2007).

As can be seen in Fig. 3.9 the submarine model is attached to a sting, and the sting is attached to two struts. In addition to the version shown in Fig. 3.9 there is an alternative arrangement whereby a sword mount connected to the struts is directly attached to the dynamometer inside the model through the sail.

3.4.2 Computational Fluid Dynamics

There are a number of Computational Fluid Dynamics (CFD) approaches which can be used to predict the values of the various manoeuvring coefficients. CFD is a fast

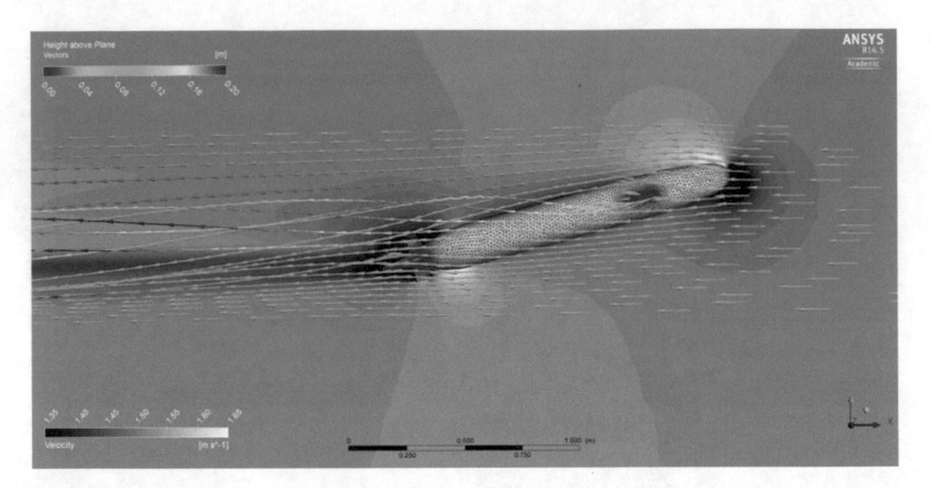

Fig. 3.10 CFD model of submarine set at a drift angle (courtesy of the Australian Maritime College)

moving field, and it is not the intention of the current text to attempt to cover the latest developments in the field. However the application of such techniques to prediction of the values of the coefficients will be covered briefly.

The usual approach is to predict the hydrodynamic coefficients required for equations representing the hydrodynamic forces and moments, such as Eqs. 3.7–3.12, by simulating a captive model experiment. An operating propulsor is required, as this influences the flow over the stern of the submarine. However this can be simplified, as it is only necessary to represent the bulk flow over the stern of the submarine, and not the detailed flow over propeller or stator blades.

To predict the values of the coefficients which represent the forces and moments as functions of sway velocity, a sway velocity can be imposed by setting the submarine model at a drift angle, as shown in Fig. 3.10. Simulation runs are completed for a range of different values of sway velocity. The assumption of symmetry is not possible, and in addition the CFD domain size will need to be greater, due to blockage, than when considering a zero drift angle. Thus, the computational effort is increased, however this is still a fairly straightforward exercise.

A similar procedure can be used to predict the coefficients which represent the forces and moments as functions of appendage angles.

In addition, it is relatively straightforward to set up a CFD simulation for the case where there are both sway and heave velocities, thus making it possible to predict the values of the coefficients which represent the forces and moments as functions of both these motion parameters. This is a particularly difficult captive experiment to perform, so the use of CFD for this purpose is a considerable advantage.

However, to predict the values of the coefficients which represent the forces and moments as functions of rotational velocities, pitch and yaw, it is necessary to

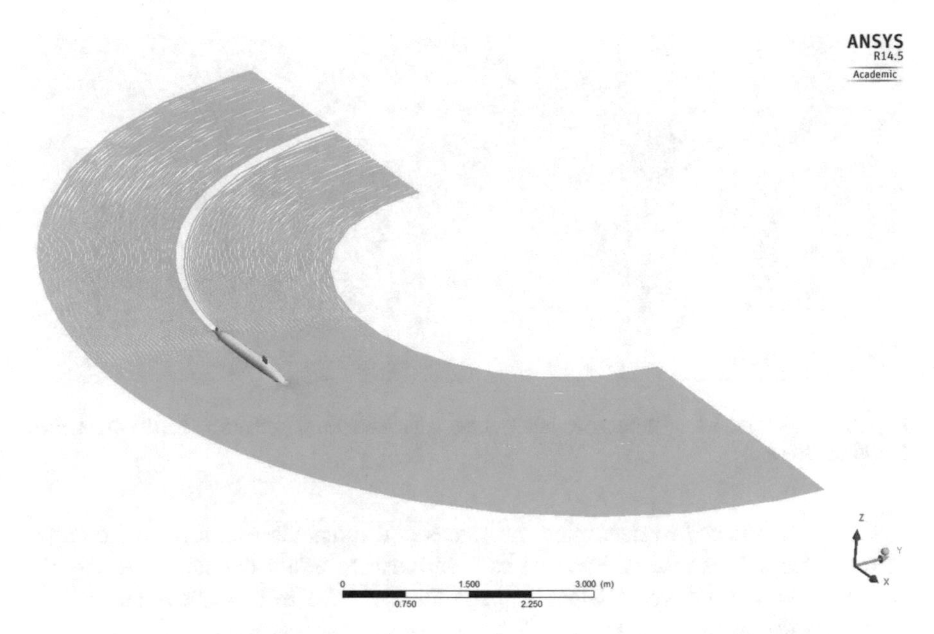

Fig. 3.11 CFD model of submarine in a rotating frame of reference (courtesy of the Australian Maritime College)

model the hull form in a circular system. Although more complex than the linear systems required to predict the values of the other coefficients, this can be achieved, albeit with an increase in computing cost. Figure 3.11 gives an example of the domain for such a calculation. In this figure the submarine has a drift angle of zero and a non-zero yaw velocity.

The added mass coefficients are non-viscous in nature, and hence easy to obtain. Thus it is not strictly necessary to use advanced CFD techniques to predict their values. However, it is possible to make use of a numerical Planar Motion Mechanism, as shown in Fig. 3.12, to do this, if required. This technique can also be used to obtain many of the other coefficients, although it is extremely costly in computing power, and, just like the physical model experiments, care needs to be taken with frequency dependence of the coefficients.

One of the great advantages of CFD compared to physical model experiments is the ability to obtain the coefficients at full scale Reynolds numbers. However, this does further complicate the CFD approach, as the smaller wake requires a greater grid size. In addition, it is not possible to compare the results from the CFD predictions with those from captive model experiments.

Another advantage of the CFD compared to physical model experiments is the ability to carry out the predictions without the presence of the support struts which are necessary for the captive physical model tests. It is also possible to make use of CFD to predict the effects of the support struts on the physical model, and hence to correct the results of physical model tests for them.

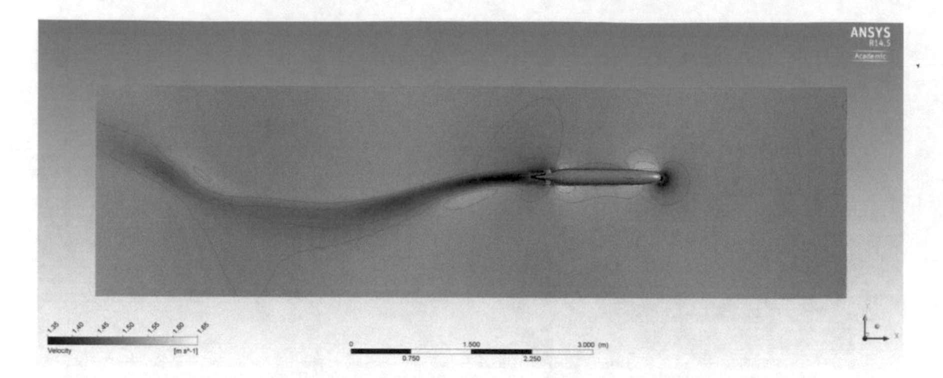

Fig. 3.12 CFD model of submarine undergoing planar motions (courtesy of the Australian Maritime College)

CFD can be used to determine the forces and moments on individual components of the submarine, such as the sail, and also to easily conduct flow visualisation. Both can be very valuable, particularly if the results show unusual or unexpected trends.

One of the big difficulties with using CFD techniques to predict the performance of a submarine is that such techniques are advancing at a very high rate. This means that the results from a prediction made today, using current techniques, may differ from those for the same hull form made only a few years ago. This is quite different to most physical model tests, which have been refined over a period of years and now remain essentially constant, at least for routine tests used to predict full scale performance. This means that it is difficult to develop empirical correlation factors, which will always be necessary to bridge the gap between even the best prediction techniques, and full scale performance.

Thus, it is important for any hydrodynamics organisation to carefully record the numerical techniques used, and preferably not to continuously change these, but only to do so in a well-documented step form.

3.4.3 Approximation Techniques

3.4.3.1 General

In the early design stage it is desirable to be able to estimate the manoeuvring characteristics of a submarine. This will be necessary long before either advanced CFD modelling or physical model experiments are commissioned. This is needed to determine the size of the appendages, for example, to give the desirable level of straight line stability. When this is required simple approximation methods for predicting the values of the linear coefficients are needed.

It is known that these approximation methods are not necessarily very accurate, as discussed in Jones et al. (2002). This means that the results from these initial predictions should be treated with care, and confirmed by either CFD or physical model experiments. It is unrealistic to expect these approximation methods to be able to accurately predict the non-linear coefficients.

Unlike the situation for surface ships, there is not a large library of publically available experimental data to draw upon.

One approach is to first determine the coefficients for the unappended hull, and then add the influence of the appendages, including the sail. Interaction between the hull and the appendages may need to be considered, although for a preliminary estimation this may not be required.

Note that, in general, the contribution to many of the manoeuvring coefficients from the hull is quite low compared with that from the sail and planes.

3.4.3.2 Hull

For a purely axisymmetric body, the side force/yawing moment due to sway motion/sway acceleration will be identical to the vertical force/pitching moment due to heave motion/heave acceleration, and the side force/yawing moment due to yaw motion/acceleration will be identical to the heave force/pitching moment due to pitch motion/acceleration. This leads to Eqs. 3.49–3.56. Note that due to the sign convention, the coefficients: N'_v and M'_w; and $N'_{\dot{v}}$ and $M'_{\dot{w}}$ will have opposite signs.

$$Y'_v = Z'_w \tag{3.49}$$

$$Y'_{\dot{v}} = Z'_{\dot{w}} \tag{3.50}$$

$$N'_v = -M'_w \tag{3.51}$$

$$N'_{\dot{v}} = -M'_{\dot{w}} \tag{3.52}$$

$$Y'_r = Z'_q \tag{3.53}$$

$$Y'_{\dot{r}} = Z'_{\dot{q}} \tag{3.54}$$

$$N'_r = M'_q \tag{3.55}$$

$$N'_{\dot{r}} = M'_{\dot{q}} \tag{3.56}$$

For an axisymmetric body all the coupling between roll moment and motions in the horizontal and vertical planes will be zero. In addition, a number of the linear terms will be zero, due to symmetry, as given in Eq. 3.57.

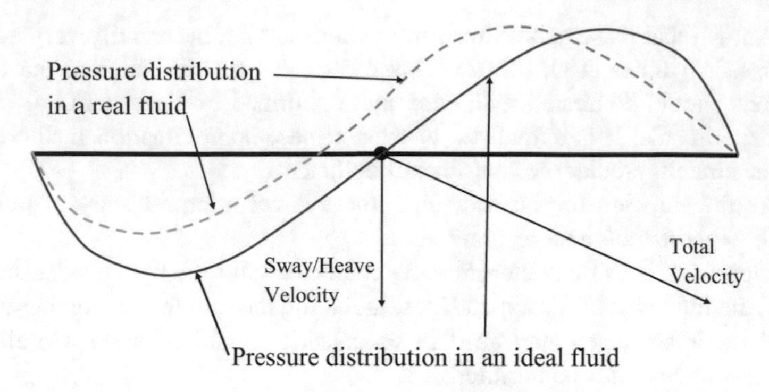

Fig. 3.13 Schematic of pressure distribution on a submarine in pure sway/heave (ideal fluid solid line, real fluid dotted line). Bow is on the right

$$Y'_{\dot{p}} = Y'_p = Y'_* = Z'_* = Z'_v = M'_* = N'_{\dot{r}} = N'_{\dot{p}} = N'_p = N'_* = 0 \tag{3.57}$$

In ideal flow the pressure distribution along the length of a symmetrical hull in pure sway or pure heave is illustrated schematically in Fig. 3.13. As can be seen, there will be no net side force as the pressure over the forward half of the body will be identical to the pressure over the aft half of the body. However there will be a moment, known as the "Munk moment". Thus, the coefficients for a hull with fore and aft symmetry in an ideal flow can be obtained by Eqs. 3.58–3.68.

$$Y'_v = Z'_w = 0 \tag{3.58}$$

$$Y'_r = Z'_q = 0 \tag{3.59}$$

$$N'_r = M'_q = 0 \tag{3.60}$$

$$N'_v = -\left(k_y + k_x\right)m' \tag{3.61}$$

$$M'_w = \left(k_z + k_x\right)m' \tag{3.62}$$

$$Y'_{\dot{v}} = -k_y m' \tag{3.63}$$

$$Z'_{\dot{w}} = -k_z m' \tag{3.64}$$

$$Y'_{\dot{r}} = N'_{\dot{v}} = 0 \tag{3.65}$$

$$Z'_{\dot{q}} = M'_{\dot{w}} = 0 \tag{3.66}$$

$$N'_{\dot{r}} = -k_z I'_{zz} \tag{3.67}$$

$$M'_{\dot{q}} = -k_y I'_{yy} \tag{3.68}$$

k_x, k_y, and k_z are the added mass coefficients for motion in the x, y and z directions respectively. I'_{yy} and I'_{zz} are the non-dimensional mass moments of inertia in pitch and yaw respectively.

It is generally considered sufficiently accurate to use the values for an ellipsoid of revolution with the same length and diameter to obtain the added mass coefficients. Note that for an axisymmetric body $k_y = k_z$ and $I'_{yy} = I'_{zz}$.

Equation 3.69 can be used to obtain the diameter of the equivalent ellipsoid of revolution, $\bar{d}$, such that the mass of the ellipsoid of revolution is the same as the mass of the submarine.

$$\bar{d} = \left(\frac{6\Delta}{\pi\rho L}\right)^{0.5} \tag{3.69}$$

Reasonable approximations of k_y (= k_z), and k_x for an ellipsoid of revolution can be estimated using Eqs. 3.70 and 3.71, obtained from results given in Korotkin (2009).

$$k_y = k_z = -0.00088\left(\frac{L}{\bar{d}}\right)^2 + 0.0245\left(\frac{L}{\bar{d}}\right) + 0.805 \tag{3.70}$$

$$k_x = -0.00047\left(\frac{L}{\bar{d}}\right)^2 + 0.0134\left(\frac{L}{\bar{d}}\right) - 0.059 \tag{3.71}$$

The mass moments of inertia of the equivalent ellipsoid are given by Eq. 3.72.

$$I_{yy} = I_{zz} = \frac{\pi\rho}{30} L\bar{d}^4\left(\left(\frac{L}{\bar{d}}\right)^2 + 1\right) \tag{3.72}$$

The mass and mass moments of inertia are non-dimensionalised as given in Eqs. 3.73 and 3.74.

$$m' = \frac{\Delta}{\frac{1}{2}\rho L^3} \tag{3.73}$$

$$I'_{yy} = \frac{I_{yy}}{\frac{1}{2}\rho L^5} \tag{3.74}$$

In a real fluid the pressure distribution is altered by the presence of viscosity, as shown schematically in Fig. 3.13 (dotted line). Thus there will be a net transverse force on a body at an angle of attack, and consequently Eqs. 3.58–3.62 are not

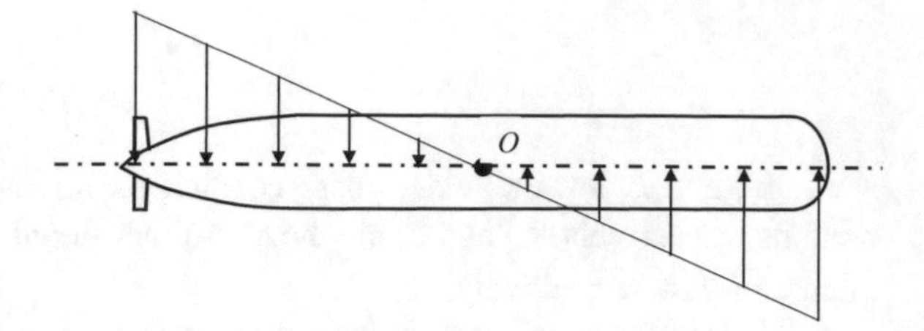

Fig. 3.14 Schematic of transverse flow on submarine undergoing rotary motion (yaw or pitch)

applicable in a real fluid. The added mass is not affected by viscosity, and hence Eqs. 3.63–3.68 are valid even in real fluid.

Expressions for Z'_w and for M'_w taken from experimental results published by Praveen and Krishnankutty (2013) are given in Eqs. 3.75 and 3.76.

$$Z'_w = \left[0.5\left(\frac{L}{D}\right) - 11\right] \times 10^{-3} \tag{3.75}$$

$$M'_w = \left[-\left(\frac{L}{D}\right) + 20\right] \times 10^{-3} \tag{3.76}$$

Note that: $Y'_v = Z'_w$ (Eq. 3.49) and $N'_v = -M'_w$ (Eq. 3.51).

Empirical expressions for the contributions to the coefficients of force and moment due to rotary motion (Y'_r Z'_q, N'_r, and M'_q) from the hull are not available. However, the total contribution of the hull to these coefficients is likely to be very small, as they are dominated by the influence of the appendages, particularly those at the stern. This is because the local transverse velocity is much larger at the ends of the boat than over the main part of the body, as illustrated by Fig. 3.14. Thus, Eqs. 3.59 and 3.60 are reasonable approximations, even in a real fluid.

3.4.3.3 Fixed Appendages

Fixed appendages include the sail, the bow planes, the stern planes, the rudders, and the duct, where fitted. The force on each of these will be caused by the local angle of attack to the flow over them. For simplicity they can be treated as lifting surfaces with the lift and drag obtained from Eqs. 3.77 and 3.78.

$$Lift = \frac{1}{2}\rho V^2 S_a \alpha C_{L_\alpha} \tag{3.77}$$

$$Drag = \frac{1}{2}\rho V^2 S_a \alpha C_{D_\alpha} \tag{3.78}$$

where C_{L_α} is the non-dimensional slope of lift as a function of angle of attack, C_{D_α} is the non-dimensional slope of drag as a function of angle of attack, α is the angle of attack, V is the velocity and S_a is the plan form area of the lifting surface. Note that for appendages affected by the presence of the hull ahead of them, V should be modified, as discussed in Sect. 5.1.

The transverse force on the body is given by Eq. 3.79.

$$Transverse\,force = Lift\cos\alpha + Drag\sin\alpha \tag{3.79}$$

For horizontal motions this translates into the rate of change of side force as a function of sway velocity, $Y_{v_{app}}$, and for vertical motions into the rate of change of vertical force as a function of heave velocity, $Z_{w_{app}}$, in Eqs. 3.80 and 3.81.

$$Y_{v_{app}} = -\frac{1}{2}\rho V S_a (C_{L_\alpha} + C_D) \tag{3.80}$$

$$Z_{w_{app}} = -\frac{1}{2}\rho V S_a (C_{L_\alpha} + C_D) \tag{3.81}$$

C_D is the non-dimensional drag at zero angle of attack. Note that Eqs. 3.80 and 3.81 are dimensional. These can be non-dimensionalised in the usual way, based on the length squared of the submarine ($Y'_{v_{app}} = Y_{v_{app}} / \left(\frac{1}{2}\rho V L^2\right)$, and $Z'_{v_{app}} = Z_{v_{app}} / \left(\frac{1}{2}\rho V L^2\right)$).

In each case it is assumed that the lift slope is that generating lift in the desired direction. In other words, for Eq. 3.80 the appendages of interest are the sail and the rudder (fixed), and the relevant lift slopes are used, and for Eq. 3.81 the appendages of interest are the bow and stern planes (fixed). If a stern configuration other than a cruciform one is used, then the components of the lift slopes in the relative directions are required. Note also that if a propulsor duct is fitted then the forces due to this in the relevant plane should be included.

Thus, to obtain the contribution of the appendages to the manoeuvring coefficients, Y_v and Z_w, it is necessary to know the lift slopes and the drag at zero lift of these appendages. Various empirical methods exist for predicting these in the absence of experimental or numerical data, including those given by references such as: Lyons and Bisgood (1950), Whicker and Fehiner (1958), Abbott and Von Doenhoff (1960) and Molland and Turnock (2007).

It is not intended to repeat all these various empirical methods here, however it should be noted that as the appendages (particularly those at the stern) contribute significantly to the manoeuvring coefficients it is important to be able to predict the lift slopes on these as accurately as possible. Organisations which have experience with particular designs of appendages should take advantage of data for these appendages, which may be proprietary, and it is recommended that such information be validated wherever possible using either model experiments or advanced CFD.

Note also that care needs to be taken when estimating the lift slope for stern appendages, as these are influenced by the presence of the hull ahead of them, and they may operate in decelerating flow due to the half tail cone angle. This is discussed more fully in Dempsey (1997), Mackay (2001, 2003) and Bettle (2014).

The effect of the presence of the hull on the appendages is discussed by Bettle (2014) where efficiency factors are obtained for each appendage. The efficiency factor is defined as the ratio between the hydrodynamic load produced by the appendage attached to the hull, and that of an equivalent appendage in isolation (mounted on a ground board).

The sail efficiency factor given by Bettle (2014), taken from Pitts et al. (1957), is given in Eq. 3.82.

$$k_s = \left(1 + \frac{b_{de}}{b_e}\right)^2 \tag{3.82}$$

where b_{de} and b_e are the distances from an effective hull centerline to the top of the deck casing and to the tip of the sail, respectively. The effective hull centerline is moved above the actual hull centerline by the distance from the top of the bare hull to the top of the deck casing.

Note that the sail efficiency factor is greater than one, which is due to the increased flow over the sail caused by the three dimensional effect of the hull compared to the flow over a ground-board.

The sail plane efficiency factor given by Bettle (2014) is estimated from Eq. 3.83.

$$k_{sp} = 1 - \left(\frac{b_d}{b_{sp}}\right)^2 \tag{3.83}$$

where b_d and b_{sp} are the deck and sail plane heights respectively, both measured from the hull centerline.

The stern plane efficiency factor is given by Bettle (2014) in Eqs. 3.84 and 3.85. It is noted that this may be conservative, however at the time of writing Bettle recommended using the value from Eqs. 3.84 and 3.85. This also applies to the rudder.

$$k_{WB} = 1 - \frac{0.2556}{\left(\frac{b}{r_M}\right)^2}\sqrt{\left(\frac{b}{r_M}\right)^2 - 0.1612} - 0.6366\sin^{-1}\left(\frac{0.4015}{\frac{b}{r_M}}\right) \tag{3.84}$$

for: $0.4015 < \frac{b}{r_M} < 0.734$ and $1.426 < \frac{b}{r_M} < \infty$

and

$$k_{WB} = -0.3644 + 1.2380\frac{b}{r_M} - 0.3728\left(\frac{b}{r_M}\right)^2 \tag{3.85}$$

for $0.734 \leq \frac{b}{r_M} \leq 1.426$

where b is the stern plane fin span from the hull centreline, and r_M is the maximum hull radius.

The contribution of each appendage to the moment coefficients is given in Eqs. 3.86 and 3.87 by making use of the lever from the origin to the centre of pressure of the appendage.

$$N_{v_{app}} = l_{app} \times Y_{v_{app}} \tag{3.86}$$

$$M_{w_{app}} = -l_{app} \times Z_{w_{app}} \tag{3.87}$$

where l_{app} is the horizontal coordinate of the centre of pressure of the appendage. $N_{v_{app}}$ and $M_{w_{app}}$ will have opposite signs due to the sign convention.

Equations 3.86 and 3.87 are dimensional and can be non-dimensionalised in the usual way, based on the length cubed of the submarine, ($N'_{v_{app}} = N_{v_{app}} / \left(\frac{1}{2}\rho V L^3\right)$, and $M'_{v_{app}} = M_{v_{app}} / \left(\frac{1}{2}\rho V L^3\right)$).

The contributions of the appendages to the rotary coefficients are given in Eqs. 3.88–3.91.

$$Y_{r_{app}} = l_{app} \times Y_{v_{app}} \tag{3.88}$$

$$Z_{q_{app}} = l_{app} \times Z_{w_{app}} \tag{3.89}$$

$$N_{r_{app}} = l_{app}^2 \times Y_{v_{app}} \tag{3.90}$$

$$M_{q_{app}} = l_{app}^2 \times Z_{w_{app}} \tag{3.91}$$

The added masses of the appendages can be obtained by assuming that each appendage is a flat plate, although a more sophisticated approach may be required for a blended sail. However, it is important to note that values of the added masses do not have a major influence on the submarine manoeuvres, as the added masses are added to the actual masses to obtain the total coefficients used for prediction of the submarine manoeuvring characteristics.

The added mass of a flat plate can be estimated using Eq. 3.92 taken from Dong (1978).

$$m_{added} = \pi K_a \rho \frac{a^2}{4} b \tag{3.92}$$

where a is the chord of the flat plate and b is its span. K_a is a coefficient which depends on the aspect ratio, as given in Table 3.3.

For flat plates with a ground-board the effective span is double the geometric span. Thus, for rudders and planes with geometric aspect ratios greater than 1.5, the added mass can be obtained from Eq. 3.93.

Table 3.3 Values of K_a from Dong (1978)

b/a	K_a
1.0	0.478
1.5	0.680
2.0	0.840
2.5	0.953
3.0	1.000
3.5	1.000
4.0	1.000

$$m_{added} = \pi\rho\frac{a^2}{4}b \tag{3.93}$$

If alternative methods are available for predicting the added mass they can be used to obtain m_{added} for the appendages.

The contributions of the appendage to the values of the linear sway and heave added mass coefficients are given by Eqs. 3.94 and 3.95.

$$Y'_{\dot{v}_{app}} = -\frac{m_{added}}{\frac{1}{2}\rho L^3} \tag{3.94}$$

$$Z'_{\dot{w}_{app}} = -\frac{m_{added}}{\frac{1}{2}\rho L^3} \tag{3.95}$$

where the values of m_{added} used are for the appropriate direction—the sail and rudders contribute to the value for $Y'_{\dot{v}_{app}}$ and the planes to the values for $Z'_{\dot{w}_{app}}$.

The contributions of the appendages to the other coefficients are given in Eqs. 3.96–3.101 where l_{app} is the horizontal coordinate of the centre of added mass of the appendage.

$$N'_{\dot{v}_{app}} = \frac{l_{app} \times Y'_{\dot{v}_{app}}}{L} \tag{3.96}$$

$$M'_{\dot{w}_{app}} = -\frac{l_{app} \times Z'_{\dot{w}_{app}}}{L} \tag{3.97}$$

$$Y'_{\dot{r}_{app}} = \frac{l_{app} \times Y'_{\dot{v}_{app}}}{L} \tag{3.98}$$

$$Z'_{\dot{q}_{app}} = \frac{l_{app} \times Z'_{\dot{w}_{app}}}{L} \tag{3.99}$$

$$N'_{\dot{r}_{app}} = \frac{l^2_{app} \times Y'_{\dot{v}_{app}}}{L^2} \tag{3.100}$$

$$M'_{\dot{q}_{app}} = \frac{l_{app}^2 \times Z'_{\dot{w}_{app}}}{L^2} \tag{3.101}$$

3.4.3.4 Propeller

An open propeller will generate a side force when it is at an angle to the flow. This side force will contribute to the manoeuvring coefficients in the same way as a fixed fin located at the longitudinal position of the propeller.

Thus, in order to calculate the contribution of the propeller to the manoeuvring coefficients, it is necessary to know the rate of change of side force as a function of inflow angle (analogous to the lift slope for a fixed fin).

There have been a number of approaches to develop empirical methods to predict the propeller side force when operating at an angle of drift including: Harris (1918), Ribner (1943) and Gutsche (1975). Of these, the Harris method is the easiest to use at the early design stage, as given by Eq. 3.102, taken from UCL (undated). Note that this is based on experiments using aircraft propellers.

$$Z'_{w_{Prop}} = Y'_{v_{Prop}} = -\frac{4.24D^2}{JL^2}\left[K_Q - \frac{J}{2}\frac{dK_Q}{dJ}\right] \tag{3.102}$$

Both the Ribner and the Gutsche method require detailed knowledge about the geometry of the propeller. Note that a good description of the Ribner method is given in Bonci (2014).

A simplification of the Gutsche method, from Dubbioso et al. (2013), is given in Eq. 3.103.

$$K_{Ty} = \left[2K_{Q_{(J=0)}} - J\frac{dK_Q}{dJ}\right] J\tan\psi \tag{3.103}$$

In Eq. 3.103 K_{Ty} is the non-dimensional side force and $K_{Q_{(J=0)}}$ is the value of the torque coefficient for $J = 0$.

Note that the vertical force due to pitch angle can be obtained from Eq. 3.103 by substituting the pitch angle, θ, for the drift angle, ψ.

The values of the hydrodynamic normal coefficients can therefore be obtained from Eq. 3.104.

$$Z'_{w_{Prop}} = Y'_{v_{Prop}} = -\frac{2(1-w)^2 D^2\left[2K_{Q_{(J=0)}} - J\frac{dK_Q}{dJ}\right]}{JL^2} \tag{3.104}$$

It is recommended to use Eq. 3.104 at the initial design stage.

Dubbioso et al. (2013) gives results for an open propeller where the Ribner and the Gutsche methods are compared with those from Blade Element Momentum Theory and CFD. This showed that the Ribner and Gutsche methods give similar results, which compare reasonably well with those from CFD. The Harris method also gives results which are acceptable for the initial design stage.

At a later stage in the design the use of CFD to more accurately determine the side force is recommended. Such work has been conducted for surface ships by a number of researchers including: Ortolani et al. (2015), Broglia et al. (2015), Dubbioso et al. (2017) and Sun et al. (2018).

3.4.3.5 Control Surfaces

The effect of control surfaces on the forces and moments on the submarine can be modelled using linear coefficients, as given in Eqs. 3.8–3.12. The values of these coefficients can be estimated using Eqs. 3.105–3.110 if the lift slope of the active part of the control surface, C_{L_δ}, is known. The subscripts "B", "R" and "S" refer to the bow planes, the rudder and the stern planes respectively. Note that if a stern form other than a cruciform one is used then these coefficients will differ, as discussed in Sect. 6.4.

$$Y_{\delta R} = \frac{1}{2}\rho V_R^2 A_{plan_{rudder}} C_{L_{\delta R}} \tag{3.105}$$

$$N_{\delta R} = \frac{1}{2}\rho V_R^2 A_{plan_{rudder}} x_{rudder} C_{L_{\delta R}} \tag{3.106}$$

$$Z_{\delta B} = -\frac{1}{2}\rho V_B^2 A_{plan_{bow}} C_{L_{\delta B}} \tag{3.107}$$

$$M_{\delta B} = -\frac{1}{2}\rho V_B^2 A_{plan_{bow}} x_{bow} C_{L_{\delta B}} \tag{3.108}$$

$$Z_{\delta S} = -\frac{1}{2}\rho V_S^2 A_{plan_{stern}} C_{L_{\delta S}} \tag{3.109}$$

$$M_{\delta S} = -\frac{1}{2}\rho V_S^2 A_{plan_{stern}} x_{stern} C_{L_{\delta S}} \tag{3.110}$$

where V_R, V_B, and V_S are the local velocities at the rudder, the bow planes and the stern planes respectively. Although the local velocity at the bow planes is likely to be very close to the velocity of the submarine, the local velocity at the rudder and stern planes will be influenced by the presence of the hull and the propulsor (see Sect. 5.1).

Alternatively, an efficiency factor, as discussed in Sect. 3.4.3.3 can be applied.

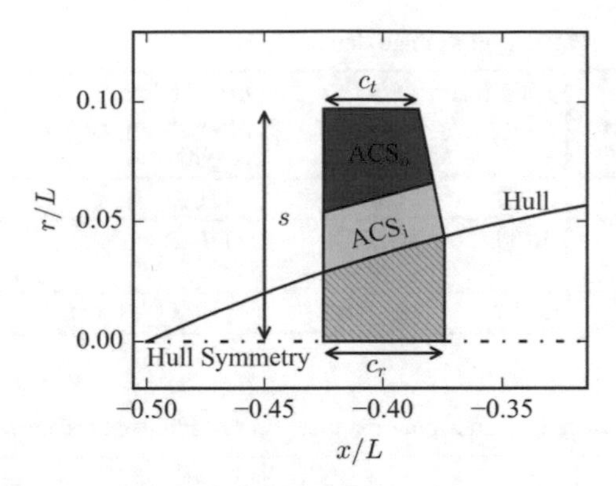

Fig. 3.15 ACS geometry used by Pook et al. (2017) (Red is moveable outer section labelled ACS$_o$. Grey is inner headbox labelled ACS$_i$. Hatched is hull area used in derivative estimation. Mirrored ACS used in derivate calculations not shown $c_t/L = 0.040$, $c_r/L = 0.060$, $s/L = 0.097$)

It should be noted that Pook et al. (2017) found that the difference between CFD and an estimate based on the method of Lyons and Bisgood (1950) was approximately 9% for the aft control surfaces on an X-form configuration. However in this case the aft control surfaces were somewhat outboard of the boundary layer, as shown in Fig. 3.15.

3.4.3.6 Predictions for Suboff

The above method was applied to the DARPA standard submarine hull form, Suboff (Sect. 1.3.4). Details of this hull form are given by Groves et al. (1989) and Roddy (1990). The principal particulars of this model are given in Table 3.4. Note that these dimensions are model scale. The length between particulars has been used for non-dimensionalising purposes. The origin used in the experiments was 2.013 m aft of the bow. This was also used for the predictions.

A summary of the appendages is given in Table 3.5, along with the estimated values of lift slope coefficient, and drag coefficient, using empirical methods. Note that the model did not have bow planes, and that the stern planes were identical to the rudders.

Table 3.4 Principal particulars of DARPA suboff model (from Roddy 1990)

Parameter	Symbol	Value
Length overall	Loa	4.356 m
Length between perpendiculars	Lbp	4.261 m
Diameter	D	0.508 m
Fineness ratio	L/D	8.575
Mass	m	705.9 kg

Table 3.5 Summary of model appendage details

	Average chord (m)	Span (m)	Plan area (m^2)	Longitudinal position (m) (from origin)	C_{L_α}/ radian	C_D
Sail	0.368	0.222	0.0794	1.022	1.4	0.010
Stern plane	0.183	0.134	0.0248	−1.902	1.8	0.019
Rudder	0.183	0.134	0.0248	−1.902	1.8	0.019

Table 3.6 Summary of non-dimensional manoeuvring coefficients for horizontal plane

Coefficient	Hull		Hull + Sail		Hull + Rudders	
$\times 10^{-3}$	Prediction	Expt	Prediction	Expt	Prediction	Expt
Y'_v	−6.71	−5.95	−22.11	−23.01	−9.82	−10.49
N'_v	−11.42	−12.8	−17.79	−15.53	−10.03	−11.25
Y'_r	0	1.81	−6.37	−0.02	1.39	6.32
N'_r	0	−1.60	−1.53	−2.38	−0.62	−3.06
$Y'_{\dot{v}}$	−17.20	−13.3	−17.54	−15.04	−17.32	−14.71
$N'_{\dot{v}}$	0	0.20	−0.08	0.01	0.05	0.42
$Y'_{\dot{r}}$	0	0.06	−0.08	−0.20	0.05	0.47
$N'_{\dot{r}}$	−3.49	−0.68	−3.51	−0.71	−3.51	−0.74

Table 3.7 Summary of non-dimensional manoeuvring coefficients for vertical plane

Coefficient	Hull		Hull + Stern Planes	
$\times 10^{-3}$	Prediction	Expt	Prediction	Expt
Z'_w	−6.71	−5.95	−9.82	−10.49
M'_w	11.42	12.8	10.03	11.25
Z'_q	0	1.81	1.39	6.32
M'_q	0	−1.60	−0.62	−3.06
$Z'_{\dot{w}}$	−17.20	−13.3	−17.32	−14.71
$M'_{\dot{w}}$	0	−0.20	−0.05	−0.42
$Z'_{\dot{q}}$	0	0.06	0.05	0.47
$M'_{\dot{q}}$	−3.49	−0.68	−3.51	−0.74

The predicted and measured results are given in Table 3.6 for the coefficients in the horizontal plane, and Table 3.7 for those in the vertical plane.

As Suboff is an axisymmetric body, the equivalent results for the bare hull are the same for the vertical and horizontal plane manoeuvres, with the exception of: N'_v and M'_w; and $N'_{\dot{v}}$ and $M'_{\dot{w}}$ which have opposite signs due to the sign convention. Equally, as the rudders are identical to the stern planes, the same applies to the coefficients for the cases with the rudders and the stern planes. However, the sail only affects the coefficients in the horizontal plane.

Although the empirical method gives reasonable results for some of the coefficients, for others the agreement is not good at all. This was noted by Jones et al. (2002), where a comparison was made with a range of different empirical methods.

3.4.3.7 Discussion

As noted above, existing empirical methods are not particularly good techniques for predicting the manoeuvring coefficients of a submarine.

However, such techniques can be used to modify the coefficients for a submarine hull form which is close to the new design. The effect of changes in the design can be taken into account, where such information is available. For example, if the sail is the same relative size, but located in a different longitudinal position, then the value of the coefficient Y_v will be unchanged, but the value of N_v will be changed. Its change will depend on the new location of the sail compared to the old one.

Also, if either experimental or CFD data is available for an existing submarine with a similar appendage configuration, then the lift slope and drag of that appendage configuration can be deduced from these results, and used to predict the manoeuvring coefficients for the new boat.

Either way, the limitations of such empirical techniques must be kept in mind, and either experiments or CFD conducted on any proposed hull form as early in the design stage as possible.

3.5 Alternative Approach to Simulation of Manoeuvring

The basic assumption made for the approach discussed in Sect. 3.3 is that the total hydrodynamic force and moment on a manoeuvring submarine can be obtained by knowing the motion parameters, appendage angles, and propulsor rpm, at that point in time. Any effects due to earlier motions are not included.

However, it is known that when a submarine is manoeuvring vortices are shed from its hull and sail as shown in Fig. 3.16. These will affect the pressure around the downstream parts of the hull, dependent on the relative position of the vortex and the hull. If a submarine is manoeuvring, both the position of the vortex and its strength will depend on previous motion, and not only on the current motion, as assumed in the approach discussed in Sect. 3.3.

Thus an approach which takes into account past motions, originally pioneered by Lloyd (1983), is to calculate the hydrodynamic forces and moments on a submarine at any instant in time, based on the flow field around the submarine. The flow field, and resulting forces and moments, are updated at successive intervals of time and used in a simulation process based on Eqs. 3.1–3.6.

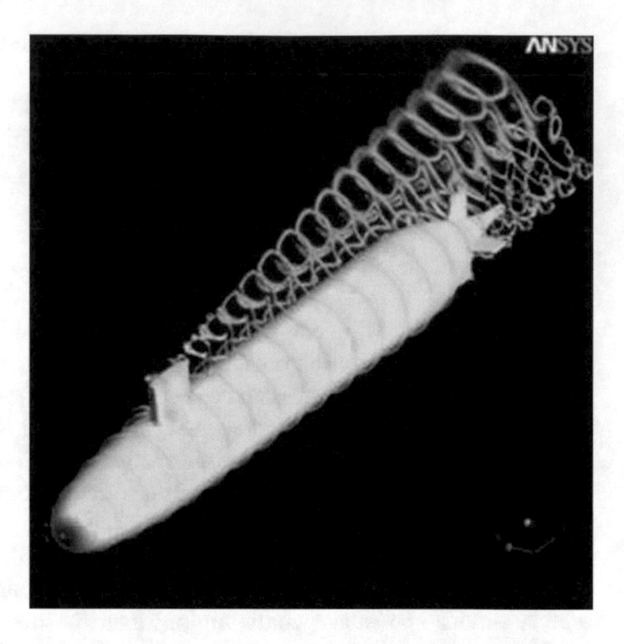

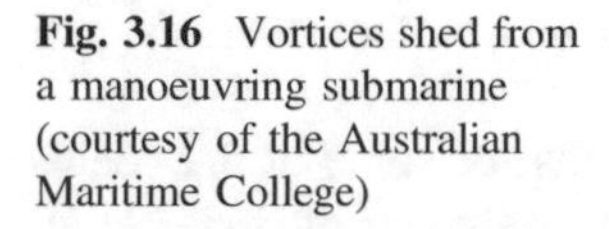
Fig. 3.16 Vortices shed from a manoeuvring submarine (courtesy of the Australian Maritime College)

To make use of this concept it is necessary to be able to determine where vortices originate from, and the strength of these vortices as functions of the motion. Then, the relative location of the vortices, along with the strength of each, can be determined at each stage in the manoeuvre. Consequently, the resulting effect on the submarine can be calculated and hence the hydrodynamic force and motion determined.

To do this, Lloyd used a flow model which employed a mixture of empirical methods and classical hydrodynamic techniques, based on an extensive range of experiments using flow visualisation to obtain information about the generation, and propagation of vortices around the hull and appendages (Lloyd and Campbell 1986). The effect of these vortices on the flow around the stern appendages was also included. The work was subsequently extended by Tinker (1988) and Ward (1992). Other work using this approach was conducted by Mendenhall and Perkins (1985) and Landrini et al. (1993).

The vortex based approach is computationally more extensive than that using hydrodynamic coefficients, discussed in Sect. 3.3, however with the computers available today (2018) it is a practical tool for many applications.

More recently, making use of extensive computational facilities and developments in CFD, it has been possible to compute the hydrodynamic forces and moments in the time domain based on the flow field around the submarine without recourse to empirical techniques for the prediction of the magnitude and propagation of vortices. These forces and moments are then used, together with a simulation process using Eqs. 3.1–3.6, to determine the motions of the submarine in the time domain.

CFD in the time domain takes considerably longer to run than the empirically based method developed by Lloyd. However, it promises to be an effective approach in the future, once computational facilities have been improved and developments in advanced CFD, with application to this field, have been achieved.

However, at present (2018), a simulation based on the calculation of the instantaneous hydrodynamic forces and moments for each time step using CFD is not practical for most submarine applications.

3.6 Manoeuvring in the Horizontal Plane

3.6.1 Turning

3.6.1.1 First Phase of a Turn

When a transverse force is applied near to the stern the submarine will experience a sway force, a roll moment and a yaw moment.

The initial consequence of the force and moments is to generate accelerations in sway, roll and yaw. At this stage in the turn the sway, roll and yaw velocities will be negligibly small, as will the heel angle. Since the deceleration due to the additional drag on the rudder will be negligibly small the resulting equations for sway, roll and yaw can be approximated as Eqs. 3.111–3.113 respectively. This is referred to as the first phase of the turn.

Sway

$$\begin{aligned} \mathrm{m}[\dot{v} + x_G\dot{r} - z_G\dot{p}] \approx \frac{1}{2}\rho L^4\left[Y'_{\dot{r}}\dot{r} + Y'_{\dot{p}}\dot{p}\right] + \frac{1}{2}\rho L^3\left[Y'_{\dot{v}}\dot{v}\right] \\ + \frac{1}{2}\rho L^2\left[Y'_{*}u^2 + Y'_{\delta R}u^2\delta_R\right] \end{aligned} \tag{3.111}$$

Roll

$$\begin{aligned} I_{xx}\dot{p} - \dot{r}I_{zx} - \mathrm{m}[z_G\dot{v}] \approx \frac{1}{2}\rho L^5\left[K'_{\dot{p}}\dot{p} + K'_{\dot{r}}\dot{r}\right] \\ + \frac{1}{2}\rho L^4\left[K'_{\dot{v}}\dot{v}\right] + \frac{1}{2}\rho L^3\left[K'_{*}u^2 + K'_{\delta R}u^2\delta_R\right] \end{aligned} \tag{3.112}$$

Yaw

$$\begin{aligned} I_{zz}\dot{r} - \dot{p}I_{zx} + \mathrm{m}[x_G\dot{v}] \approx \frac{1}{2}\rho L^5\left[N'_{\dot{r}}\dot{r} + N'_{\dot{p}}\dot{p}\right] \\ + \frac{1}{2}\rho L^4\left[N'_{\dot{v}}\dot{v}\right] + \frac{1}{2}\rho L^3\left[N'_{*}u^2 + N'_{\delta R}u^2\delta_R\right] \end{aligned} \tag{3.113}$$

3.6.1.2 Second Phase of a Turn

Once the accelerations have been applied for a finite time, sway, roll and yaw velocities will develop, such that Eqs. 3.111–3.113 are no longer valid. The velocities will cause hydrodynamic forces which will result in the accelerations reducing with time. This is known as the second phase of the turn, and the complete equations given in Eqs. 3.1–3.12 are required to represent the motion.

3.6.1.3 Third Phase of a Turn

Once the accelerations have reduced to zero the submarine will be in a steady turn with: $\dot{u} = \dot{v} = \dot{r} = \dot{p} = p = 0$. This is referred to as the third phase of the turn, and the equations in surge, sway, roll and yaw governing this phase are Eqs. 3.114–3.117.

Note that there will be a reduction in forward velocity caused by the additional drag on the rudder/planes, and the additional drag on the hull due to the sway and yaw velocities. Thus, the propulsor will not be operating at the self-propulsion point for this speed. The additional torque on the propulsor when at the lower forward speed may cause its rotational speed to reduce, but this will depend on the characteristics of the prime mover (electric motor, diesel engine, or steam turbine).

Surge

$$\begin{aligned}
-\mathrm{m}\left[vr + x_G r^2\right] &= \frac{1}{2}\rho L^4 \left[X'_{rr} r^2\right] \\
&+ \frac{1}{2}\rho L^3 \left[X'_{uu} + X'_{vr} vr\right] + \frac{1}{2}\rho L^2 \left[X'_{vv} v^2 + X'_{\delta R \delta R} u^2 \delta_R^2\right] \\
&+ \frac{1}{2}\rho L^2 \left[a_i u^2 + b_i u u_c + c_i u_c^2\right] \\
&+ \frac{1}{2}\rho L^2 \left[X'_{vv\eta} v^2 + X'_{\delta R \delta R \eta} \delta_R^2 u^2\right] (\eta - 1)
\end{aligned} \tag{3.114}$$

Sway

$$\begin{aligned}
\mathrm{m}\left[ur - y_G r^2\right] &= \frac{1}{2}\rho L^3 \left[Y'_r ur + Y'_{|r|\delta R} u|r|\delta_R + Y'_{v|r|} \frac{v}{|v|} |v||r|\right] \\
&+ \frac{1}{2}\rho L^2 \left[Y'_* u^2 + Y'_v uv + Y'_{v|v|} v|v| + Y'_{\delta R} u^2 \delta_R\right] \\
&+ (\mathrm{W} - \mathrm{B})\sin\phi \\
&+ \frac{1}{2}\rho L^3 [Y'_{r\eta} ur](\eta - 1) \\
&+ \frac{1}{2}\rho L^2 \left[Y'_{v\eta} uv + Y'_{v|v|\eta} v|v| + Y'_{\delta R\eta} u^2 \delta_R\right] (\eta - 1)
\end{aligned} \tag{3.115}$$

Roll

$$
\begin{aligned}
r^2 I_{yz} - \mathrm{m}[z_G ur] = {} & \frac{1}{2}\rho L^4 [K'_r ur] \\
& + \frac{1}{2}\rho L^3 \left[K'_* u^2 + K'_v uv + K'_{v|v|} v|v| + K'_{\delta R} u^2 \delta_R\right] \\
& + (y_G \mathrm{W} - y_B \mathrm{B})\cos\phi - (z_G \mathrm{W} - z_B \mathrm{B})\sin\phi \\
& + \frac{1}{2}\rho L^3 K'_{*\eta} u^2 (\eta - 1)
\end{aligned}
\tag{3.116}
$$

Yaw

$$
\begin{aligned}
\mathrm{m}[x_G ur + y_G vr] = {} & \frac{1}{2}\rho L^5 \left[N'_{r|r|} r|r|\right] \\
& + \frac{1}{2}\rho L^4 \left[N'_r ur + N'_{|r|\delta R} u|r|\delta_R + N'_{|v|r} |v|r\right] \\
& + \frac{1}{2}\rho L^3 [N'_* u^2 + N'_v uv + N'_{v|v|} v|v| + N'_{\delta R} u^2 \delta_R] \\
& + (x_G \mathrm{W} - x_B \mathrm{B})\sin\phi \\
& + \frac{1}{2}\rho L^4 N'_{r\eta} ur(\eta - 1) \\
& + \frac{1}{2}\rho L^3 \left[N'_{v\eta} uv + N'_{v|v|\eta} v|v| + N'_{\delta R\eta} u^2 \delta_R\right](\eta - 1)
\end{aligned}
\tag{3.117}
$$

Assuming that the heave force or pitch moment caused by the turn is balanced by the planes, Eqs. 3.114–3.117 can be used to obtain the values in a steady state turn.

Simplifying for linear motions (i.e. small rudder angles and forward velocity unchanged), neglecting roll, assuming W = B, $x_G = x_B$, $y_G = 0$, and considering only sway and yaw, Eqs. 3.118 and 3.119 can be obtained. The non-dimensional values of m and x_G are given in Eqs. 3.120 and 3.121 respectively.

$$
0 = \frac{1}{2}\rho L^3 [Y'_r - m'] ur + \frac{1}{2}\rho L^2 [Y'_v uv + Y'_{\delta R} u^2 \delta_R] \tag{3.118}
$$

$$
0 = \frac{1}{2}\rho L^4 [N'_r - m' x'_G] ur + \frac{1}{2}\rho L^3 [N'_v uv + N'_{\delta R} u^2 \delta_R] \tag{3.119}
$$

$$
m' = \frac{\mathrm{m}}{\frac{1}{2}\rho L^3} \tag{3.120}
$$

$$
x'_G = \frac{x_G}{L} \tag{3.121}
$$

Solving the simultaneous Eqs. 3.118 and 3.119 gives Eq. 3.122 which is an expression for the turning radius, R, where $R = V/r$.

$$\frac{R}{L} = \frac{1}{\delta_R}\left[\frac{N'_v(Y'_r - m') - (N'_r - m'x'_G)Y'_v}{N'_\delta Y'_v - N'_v Y'_\delta}\right] \tag{3.122}$$

Equation 3.122 only applies to turns with small rudder angles, and where the coupling between the horizontal motions (sway and yaw) and the vertical motions (heave and pitch) can be neglected.

3.6.2 Stability in the Horizontal Plane

The stability index in the horizontal plane, G_H, is given by Eq. 3.123 (Spencer 1968).

$$G_H = 1 + \frac{N'_v(m' - Y'_r)}{N'_r Y'_v} \tag{3.123}$$

As a high degree of manoeuvrability in the horizontal plane is desirable, G_H is usually a small positive value. See Sect. 3.9 for recommended values of G_H.

3.6.3 Pivot Point

When turning in the horizontal plane a submarine will experience both sway and yaw. The resultant local transverse flow vectors when in a steady state turn are shown schematically in Fig. 3.17, which shows a submarine turning to port. As can be seen from this figure, the local sway velocity when turning to port is from the port side near the bow, and from the starboard side at the stern. There is one position along the length of the submarine where the local sway velocity is zero. This is referred to as the Pivot Point.

The magnitude of the local sway velocity at the stern is large. This will influence the performance of the stern appendages and the propulsor when the submarine is undergoing a turn.

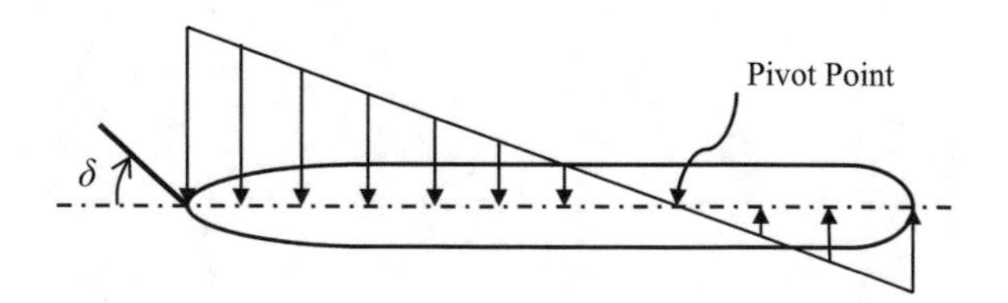

Fig. 3.17 Local transverse flow vectors in a steady turn to port

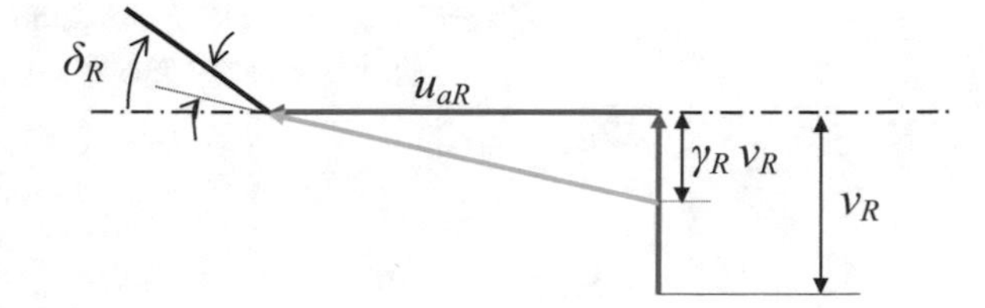

Fig. 3.18 Flow vectors at stern of submarine turning to port

3.6.4 Effective Rudder Angle

The effective angle of attack of the rudder, $\delta_{R_{eff}}$, may be significantly smaller than the actual geometrical rudder angle, and hence the yawing moment caused by the rudder will be smaller than that given by linear theory, as illustrated in Fig. 3.18. In this figure, the geometrical rudder angle is δ_R. The axial velocity at the rudder is u_{aR}. This is similar to the speed of advance, taking into account the wake, however is averaged over the rudder span, not the propeller diameter. The local sway velocity at the rudder is v_R. The actual sway velocity affecting the rudder will be less than this, due to the flow straightening effect of the presence of the submarine hull, γ_R. Thus, the effective sway velocity affecting the rudder will be $\gamma_R \mathrm{v_R}$.

Note that the value of γ_R may be close to one, as a large proportion of the rudder is quite clear of the hull, unlike the equivalent case for a surface ship, where the flow into the rudder is greatly influenced by the stern of the ship, and the propeller race. Thus, the difference between actual rudder angle, δ_R, and effective rudder angle, $\delta_{R_{eff}}$, may be greater for a submarine than for a surface ship.

The effective angle of attack of the rudder, $\delta_{R_{eff}}$, can be estimated from Eq. 3.124.

$$\delta_{R_{eff}} = \delta_R - \tan\left(\frac{\gamma_R v_R}{u_{aR}}\right) \tag{3.124}$$

The effective velocity over the rudder, $V_{R_{eff}}$, will be given by Eq. 3.125.

$$V_{R_{eff}} = \sqrt{\mathrm{u_{aR}^2} + (\gamma_R \mathrm{v_R})^2} \tag{3.125}$$

This angle and velocity can then be used together with a look-up table approach discussed in Sect. 3.3.2, to obtain values of lift and drag on the rudder, as shown in Fig. 3.19. The lift and drag can then be converted to body fixed axes using geometry, to incorporate into the left hand side of Eqs. 3.1–3.6.

Note that the same principle can also apply to an X form stern. In addition, a similar approach can be used to obtain the hydrodynamic forces and moments due to the forward and aft planes when the submarine is manoeuvring in the vertical plane.

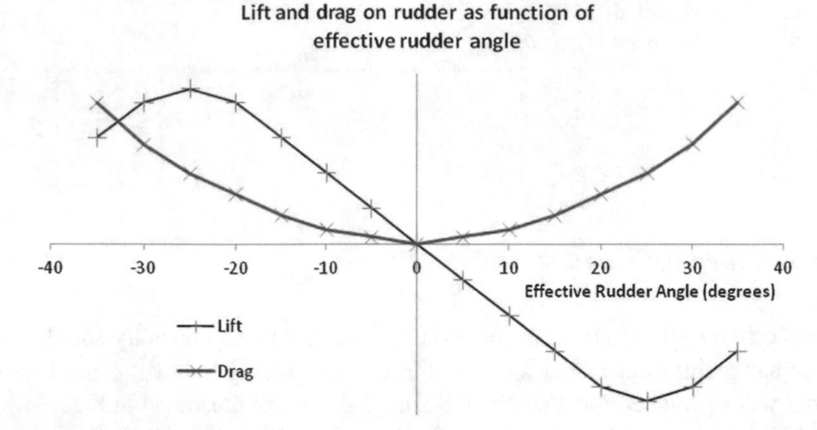

Fig. 3.19 Lift and drag on rudder as functions of $\delta_{R_{eff}}$

3.6.5 Heel in a Turn

When turning, a submarine will experience side forces on the hull and sail due to the local sway velocity. Initially this will be into the turn, as illustrated in Fig. 3.20.

This will result in a rolling moment, as illustrated in Fig. 3.21.

As can be seen from Fig. 3.21 the force on the sail acts much higher up than the forces on the rudder and the hull, resulting in a rolling moment into the turn. This is referred to as a 'snap roll' and can be a significant roll angle.

Once the submarine is in a turn the side forces on the hull and sail will depend on the local sway velocity. As the magnitude of the local sway velocity varies along the length of the submarine (Fig. 3.17) the magnitude of the sway force on the sail will depend very much on the sail's location. If the sail is located close to the Pivot Point then the force on it will be relatively small, whereas if it is located well aft of the Pivot Point then the force may be quite large.

The other factor which will influence the magnitude of the heel in the turn is the side force generated by the sail when at an angle of attack. A large foil type sail (see Sect. 6.2) will produce a greater side force than a smaller, blended sail.

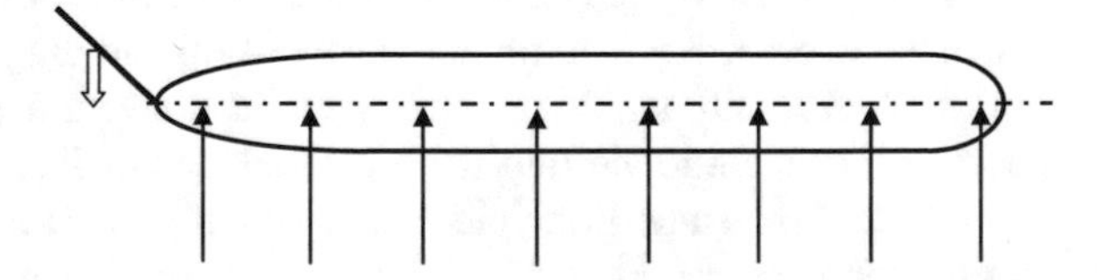

Fig. 3.20 Forces along the length of the submarine in the first phase of a turn to port

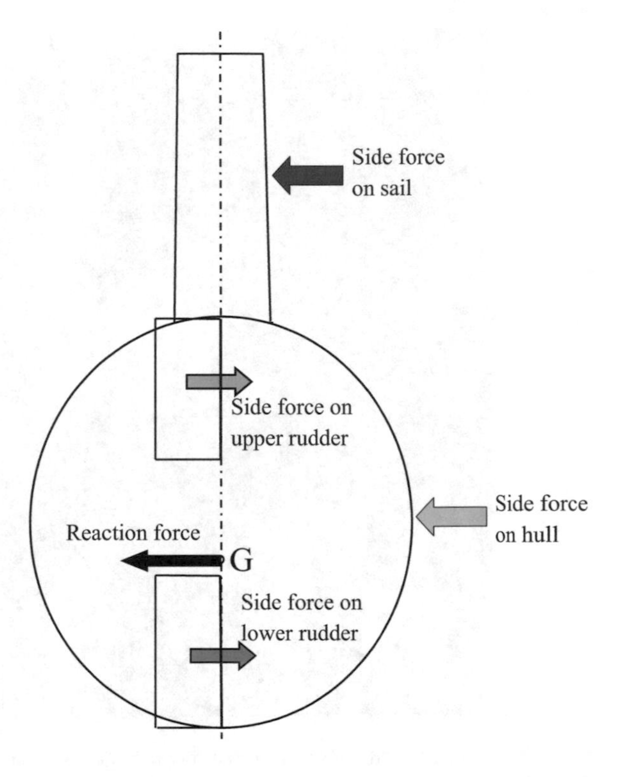

Fig. 3.21 Vertical location of forces in the first phase of a turn to port

3.6.6 Effect of Sail in a Turn

When a submarine is in a turn both the sail and the hull will generate vortices, which will interact as illustrated in Fig. 3.22. There will be a strong tip-vortex from the sail, and an opposing circulation induced by its image in the hull. This will modify the vortices shed by the top and the bottom of the hull, resulting in a force and moment in the vertical plane as discussed by Seil and Anderson (2013).

This will normally result in a downward force over the stern of the submarine. Seil and Anderson (2013) showed that the primary reason for this downward force is due to the effect of the sail on the hull-vortices, and hence the force and moment on the hull, rather than the force and moment on the sail itself.

This is represented by the coefficients: Z'_{vv}, Z'_{rr}, M'_{vv} and M'_{rr} in Eqs. 3.9 and 3.11. Note that these are non-linear terms giving the heave force and pitch moment as functions of sway velocity squared, and yaw velocity squared. This is because the heave force and pitch moment for a hull form which is symmetrical in the *x-y* plane are independent of the sign of the sway or yaw velocity.

The effect of the presence of the sail on the pressure over a manoeuvring submarine hull can be seen in Fig. 3.23 from CFD data provided by the AMC. Figure 3.22a and b show the pressure at the top and bottom of the submarine when it has zero drift angle, without and with a sail, respectively.

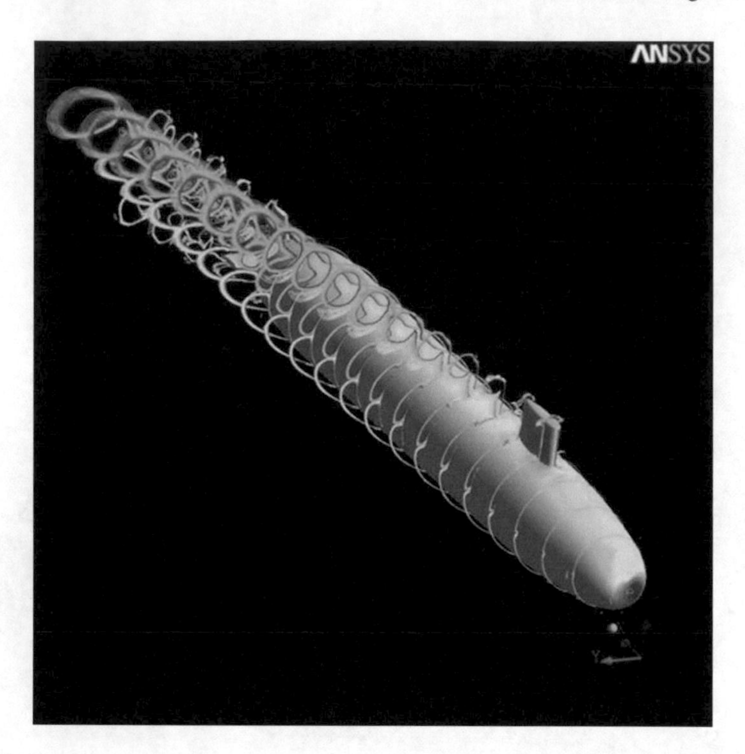

Fig. 3.22 Vortices shed from a manoeuvring submarine (courtesy of the Australian Maritime College)

As can be seen in Fig. 3.23a when there is no sail the pressure on the top and the bottom of the hull are equal. This will result in no net force in the vertical plane, and hence the values of the coefficients: Z'_* and M'_* will both be equal to zero.

The presence of the sail (Fig. 3.23b) results in a change in pressure at the location of the sail, however over the majority of the hull the pressures are similar. The difference due to the sail will result in a steady state heave force and pitch moment when the boat is travelling at $v = r = 0$. This is reflected in the values of the coefficients: Z'_* and M'_* which will be non-zero for a submarine with a sail.

Figure 3.23c and d show the pressure at the top and bottom of the submarine when it has a drift angle of 16°, without and with a sail, respectively. As expected, the pressure is reduced when the boat is at an angle to the flow. This is due to the increased velocity over the top and the bottom of the boat caused by the transverse component of the flow around the hull when it is at a drift angle.

For the case of the hull without the sail the difference in pressure between that over the top and that over the bottom is still zero, indicating that the sway velocity will not cause either a heave force or a pitch moment. Hence, the values of the coefficients: Z'_{vv}, Z'_{rr}, M'_{vv} and M'_{rr} will all be zero.

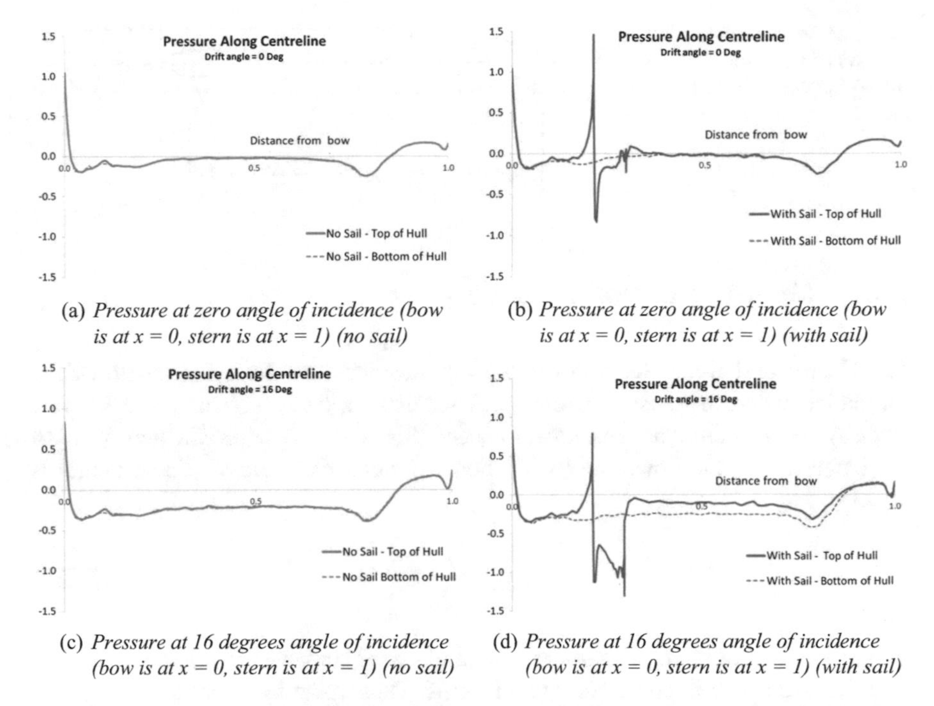

(a) *Pressure at zero angle of incidence (bow is at x = 0, stern is at x = 1) (no sail)*

(b) *Pressure at zero angle of incidence (bow is at x = 0, stern is at x = 1) (with sail)*

(c) *Pressure at 16 degrees angle of incidence (bow is at x = 0, stern is at x = 1) (no sail)*

(d) *Pressure at 16 degrees angle of incidence (bow is at x = 0, stern is at x = 1) (with sail)*

Fig. 3.23 Pressure over the length of a manoeuvring submarine with an axisymmetric hull (-courtesy of the Australian Maritime College). Bow is on the left

However, for the case of the hull with the sail (Fig. 3.23d), the difference in pressure over the top and bottom of the hull aft of the sail is quite marked. This will result in a non-zero values of the coefficients: Z'_{vv}, Z'_{rr}, M'_{vv} and M'_{rr}. As the magnitude of the negative pressure over the top of the hull aft of the sail is less than the magnitude of the negative pressure over the bottom of the hull this will result in a downward force and a trimming moment bow up, thus in this case each of these coefficients will be positive.

The result is that when turning, a submarine will normally also pitch bow up/stern down. This is known as "stern dipping". This is different to the behaviour of a purely axisymmetrical body, which will remain on level trim when turning.

Note that sail design, and/or the effect of casings, can modify the pressure distribution over the hull, and in some cases this can potentially result in the opposite effect.

It is necessary to account for this behaviour in the design of the aft appendages and any control algorithm, as clearly provision of a large rudder which can generate a tight turn is pointless if the stern plane cannot generate the required vertical force to ensure that the submarine remains on level trim.

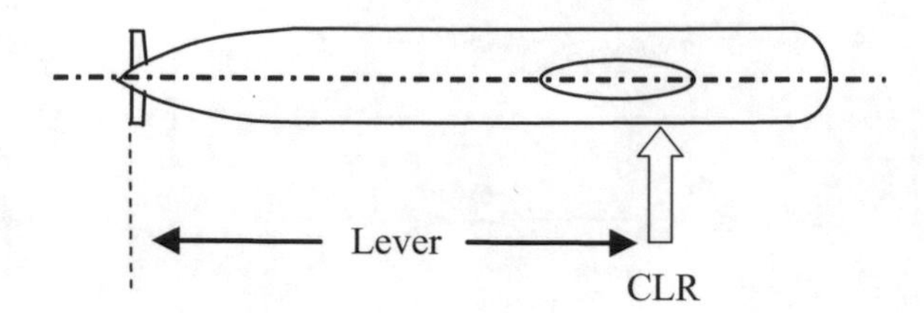

Fig. 3.24 Approximate location of CLR for submarine with forward motion

3.6.7 Centre of Lateral Resistance

The Centre of Lateral Resistance (CLR) is the position along the length of the submarine where a transverse force will result in a sway velocity, and no yaw velocity. For a submarine with forward speed this is usually approximately 1/3rd of the length aft of the bow, and its location for small drift angles (linear range) is given by Eq. 3.126.

$$x_{CLR} = \frac{N'_v}{Y'_v} L \tag{3.126}$$

The position of the CLR, x_{CLR}, is essentially fixed, however if the submarine has zero forward speed it moves close to midships. Thus, for submarines travelling very slowly it is possible that the CLR will move back towards midships.

In order to cause the submarine to turn, a transverse force is applied as far from the CLR as possible. As the CLR is forward of midships, this means applying a force as far aft as possible—which is why the rudders are placed at the stern. See Fig. 3.24.

3.7 Manoeuvring in the Vertical Plane

3.7.1 Stability in the Vertical Plane

The stability index in the vertical plane, G_V, is given by Eq. 3.127 (Spencer 1968).

$$G_V = 1 - \frac{M'_w\left(m' + Z'_q\right)}{M'_q Z'_w} \tag{3.127}$$

In the vertical plane a small degree of manoeuvrability, and a high level of stability, is desirable. This is due to the danger of large depth changes which may result in broaching when at periscope depth, or grounding/exceeding Deep Diving Depth when operating deep. This is particularly of concern when travelling at high speed. Thus, G_V is usually a large positive value. See Sect. 3.9 for recommended values of G_V.

3.7.2 Effective Plane Angles

When a submarine is pitching using its stern planes there will be a combination of heave and pitch motion, which will result in local vertical flow vectors as shown schematically in Fig. 3.25. This is analogous to the case of a submarine turning in the horizontal plane, as discussed in Sect. 3.6.

In addition, a submarine will often have operational reasons to heave without pitching, such as when operating at periscope depth. In this case the local vertical flow vectors will be as shown schematically in Fig. 3.26.

As a consequence, the flow into the bow and stern planes will not be axial, but will have a vertical component depending on the local vertical flow, in much the same way as the flow into the rudder, as discussed in Sect. 3.6.4.

Thus, following the logic given in Sect. 3.6.4, the effective angles of the bow and stern planes can be obtained from Eqs. 3.128 and 3.129.

$$\delta_{B_{eff}} = \delta_B - \tan\frac{\gamma_B w_B}{u_{aB}} \tag{3.128}$$

$$\delta_{S_{eff}} = \delta_S - \tan\frac{\gamma_S w_S}{u_{aS}} \tag{3.129}$$

The values of the flow straightening coefficients, γ_B, and γ_S, will be close to one as they are generally largely clear of the hull. In particular, the value of γ_B for midline planes or sail planes will be very close to one (see Chap. 6).

The effective velocities at the bow and stern planes, $V_{B_{eff}}$, and $V_{S_{eff}}$, respectively can be obtained from Eqs. 3.130 and 3.131.

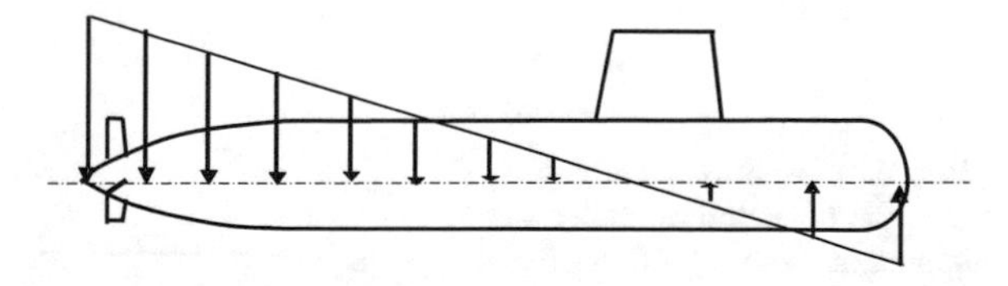

Fig. 3.25 Local vertical flow vectors in a steady pitch to dive

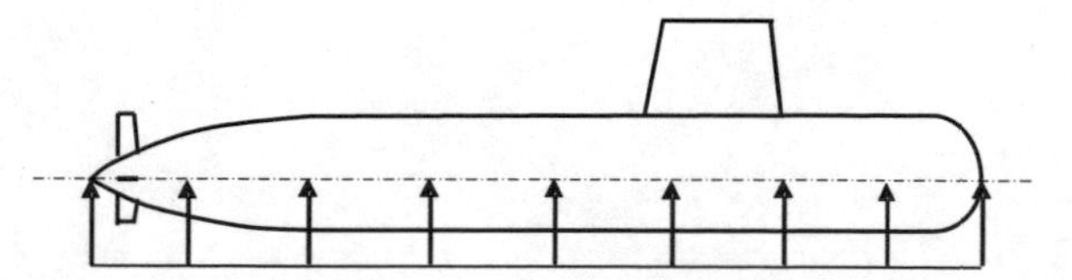

Fig. 3.26 Local vertical flow vectors in a steady heave, with no pitch

$$V_{B_{eff}} = \sqrt{u_{aB}^2 + (\gamma_B w_B)^2} \tag{3.130}$$

$$V_{S_{eff}} = \sqrt{u_{aS}^2 + (\gamma_S w_S)^2} \tag{3.131}$$

The angles and velocities can then be used together with a look-up table approach discussed in Sect. 3.3.2, to obtain values of lift and drag on the planes, which can then be converted to heave force, surge force and pitch moment using geometry, to incorporate into the left hand side of Eqs. 3.1–3.6.

3.7.3 Neutral Point

The Neutral Point is the position along the length of the submarine where a vertical force applied will cause a change in depth, but no change in pitch angle. For submarines with high forward velocity the Neutral Point is usually approximately 1/3rd of the length aft of the bow, as shown in Fig. 3.27, and its location for small pitch angles (linear range) is given by Eq. 3.132.

$$x_{NP} = -\frac{M'_w}{Z'_w} L \tag{3.132}$$

When Eq. 3.126 for the location of the Centre of Lateral Resistance (CLR) is compared with Eq. 3.132 for the Neutral Point, it can be seen that the latter has a negative sign. However, this is due to the definition of the axis system, such that for an axisymmetrical body $Y'_v = Z'_w$, but $N'_v = -M'_w$. Thus, for an axisymmetric body the longitudinal position of the Centre of Lateral Resistance is the same as the longitudinal position of the Neutral Point, as expected.

When a submarine has low forward velocity, and the heave velocity is not small compared to the surge velocity, there will be a large pitch angle, (angle of attack) as shown in Fig. 3.28, and the heave force and pitching moment will be dominated by non-linear effects.

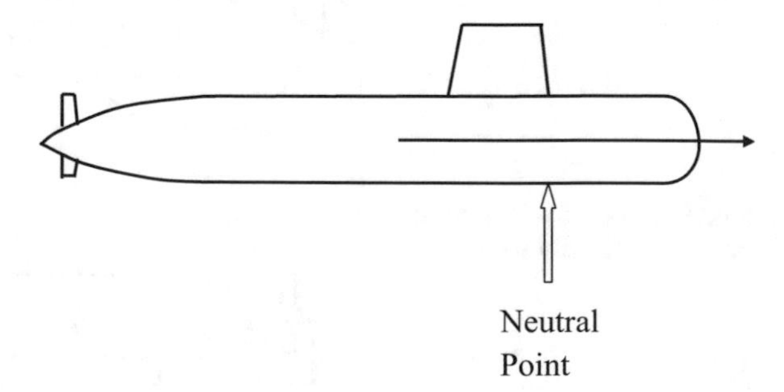

Fig. 3.27 Approximate position of neutral point for submarine with high forward velocity

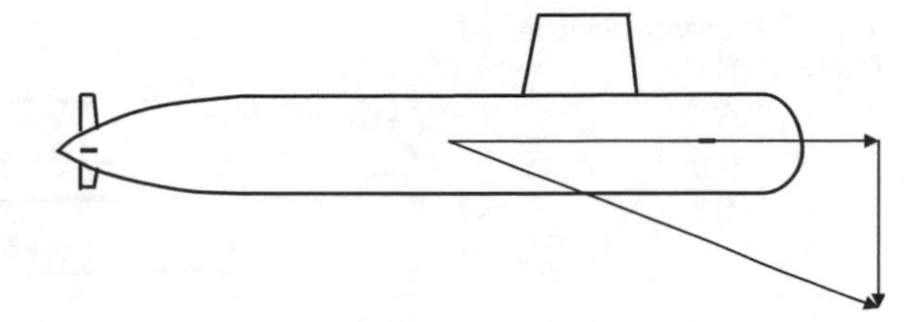

Fig. 3.28 Submarine with heave velocity and low forward velocity

In this case, the location of the Neutral Point will move aft, towards midships. To obtain its location it is necessary to make use of the expression for the hydrodynamic forces and moments incorporating the non-linear effects. Using the equations developed by Gertler and Hagen (1967), as given in Sect. 3.3, Eqs. 3.7–3.12, the position is given by Eq. 3.133.

$$x_{NP} = -\left(\frac{M'_w uw + M'_{w|w|}w|w| + M'_{|w|}u|w| + M'_{ww}w^2}{Z'_w uw + Z'_{w|w|}w|w| + Z'_{|w|}u|w| + Z'_{ww}w^2}\right)L \tag{3.133}$$

Note that for small values of w compared to u Eq. 3.133 tends to Eq. 3.132. At zero forward speed it tends to Eq. 3.134.

$$x_{NP} = -\left(\frac{M'_{w|w|}w|w| + M'_{ww}w^2}{Z'_{w|w|}w|w| + Z'_{ww}w^2}\right)L \tag{3.134}$$

For normal ahead motion Eq. 3.132 can be used. Equation 3.133 is only required for the special case where the surge velocity is low, and the heave velocity is not small, resulting in a large angle of attack, as shown in Fig. 3.28. This might be the case if the submarine is operating at low speed at periscope depth, and it takes on a significant heave velocity, perhaps due to loss of surface suction, or a change in water density.

3.7.4 Critical Point

The Critical Point is the position along the length of the submarine where a vertical force applied will cause a change in pitch angle, but no change in depth.

When a vertical force is applied at the Critical Point, there will be a trimming moment equal to the magnitude of the vertical force multiplied by the distance between the Critical Point and the Neutral Point, as shown in Fig. 3.29.

This vertical force (upward in Fig. 3.29) will cause the submarine to trim (in this case bow down). This trim will result in a vertical hydrodynamic force on the hull, which will cause a trimming moment, and a hydrostatic moment, (mgBGsinϕ) as illustrated in Fig. 3.30.

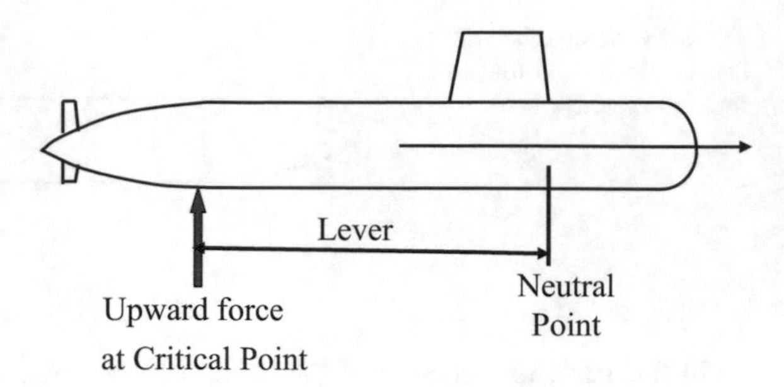

Fig. 3.29 Upward force at critical point

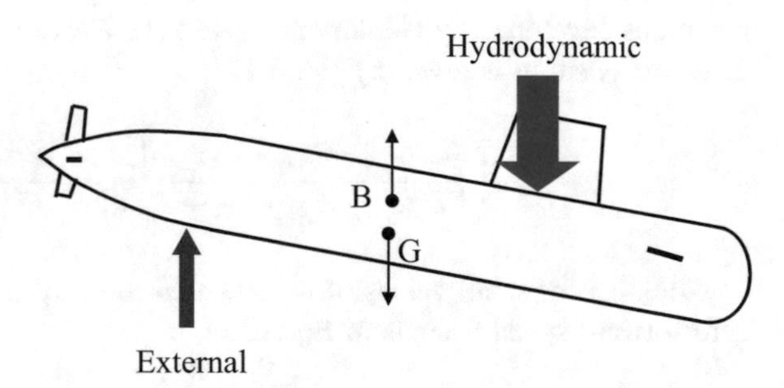

Fig. 3.30 Forces on a submarine due to upward force at critical point

By definition, if the vertical force is applied at the Critical Point, then it will be exactly balanced by the vertical hydrodynamic force, and the boat will not change its depth.

The hydrodynamic trimming moment will be proportional to velocity squared, whereas the hydrostatic trimming moment (mgBGsinϕ) will be independent of speed.

The position of the Critical Point, x_{CP}, is given by Eq. 3.135.

$$x_{CP} = \frac{2\text{mgBG}}{\rho L^2 Z'_w u^2} - L\frac{M'_w}{Z'_w} \tag{3.135}$$

As can be seen from Eq. 3.135 the location of the Critical Point is dependent on speed. For low speeds the Critical Point can become quite far aft—even astern of the stern planes.

3.7.5 *Influence of Neutral Point and Critical Point on Manoeuvring in the Vertical Plane*

The influence of the Neutral Point and the Critical Point on manoeuvring in the vertical plane will depend on the speed that the submarine is travelling.

(a) Submarine Travelling at Moderate Forward Speed

When the submarine is travelling at a moderate forward speed the position of the Neutral Point will be approximately 1/3rd of the length of the submarine aft of the bow, as discussed in Sect. 3.7.3. The position of the Critical Point will be slightly aft of this, as given by Eq. 3.135. This is illustrated in Fig. 3.31.

In this situation when an upward force is applied at the stern plane the submarine will trim bow down, due to the lever between the position of application of the upward force and the location of the Neutral Point. As the upward force is applied aft of the Critical Point, the trim angle will result in a downward hydrodynamic vertical force on the hull which is greater than the upward force at the Neutral Point, and the submarine will go downwards, as shown in Fig. 3.32.

This is the normal method of changing depth when operating at a moderate speed. The bow planes are not necessary.

(b) Submarine Travelling at Low Forward Speed

As discussed in Sect. 3.7.4, when the submarine is travelling at low forward speed the location of the Critical Point is much further aft, as given by Eq. 3.135. When the Critical Point coincides with the stern planes, as shown in Fig. 3.33, an upward force applied at the stern plane will cause the submarine to trim bow down, due to the lever between the position of application of the upward force and the location of the Neutral Point, as for the moderate speed case. However, as the upward force is now applied at the Critical Point, the downward hydrodynamic

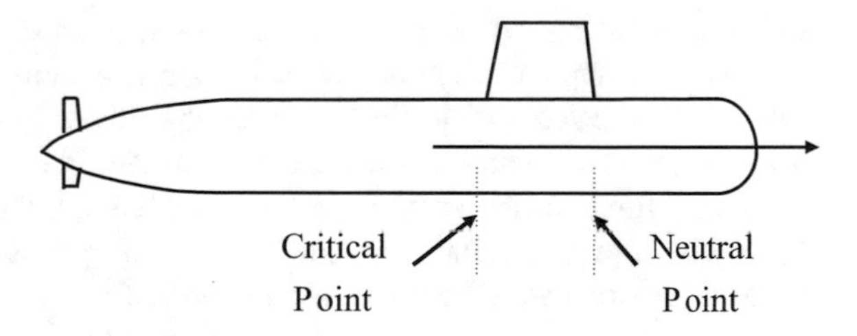

Fig. 3.31 Positions of neutral point and critical point for a submarine at moderate speed

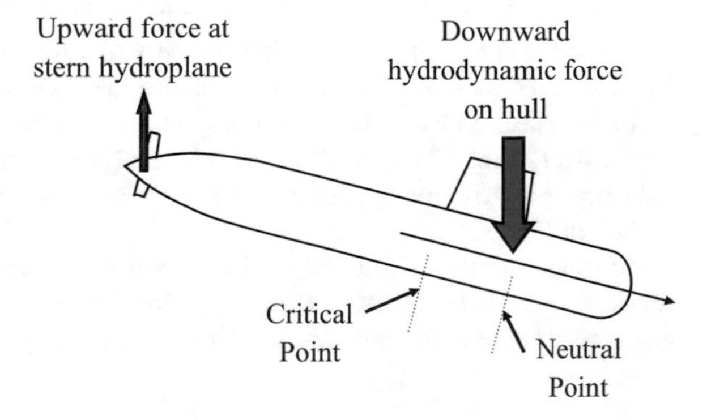

Fig. 3.32 Submarine diving when upward force at stern plane is aft of critical point

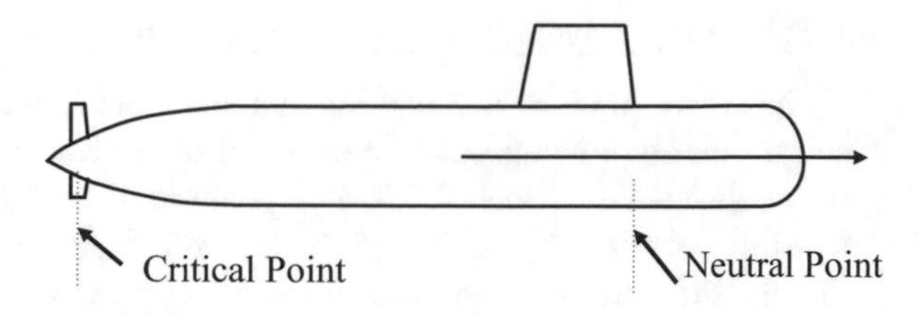

Fig. 3.33 Positions of neutral point and critical point for a submarine at low speed

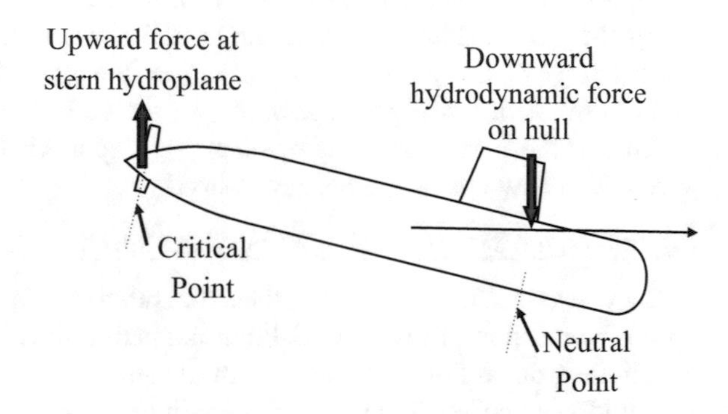

Fig. 3.34 Submarine attempting to dive using stern planes only when upward force at stern plane is at the critical point—submarine will remain at constant depth!

vertical force on the hull, due to the trim angle, will be equal to the upward force applied by the stern planes.

This will result in the boat trimming bow down, but remaining at the same depth, since the upward force on the stern plane exactly compensates for the downward hydrodynamic force on the hull, due to the trim, as shown in Fig. 3.34.

Thus, if a depth change is required when operating at low forward speeds the bow planes are required to do this. Note that if the bow planes are at the Neutral Point, then they cannot change the trim angle. They can only apply a vertical downward force to cause a dive. On the other hand, if the bow planes are forward of the Neutral Point then they can also be used to increase the trim angle, and hence increase the downward vertical hydrodynamic force on the hull, increasing the rate of descent in this situation. Further discussion of the bow planes is given in Chap. 6.

(c) Submarine Travelling at Very Low Forward Speed

When the submarine is travelling at a very low forward speed the location of the Critical Point may be aft of the stern planes, as shown in Fig. 3.35.

In this case, if an upward force is applied at the stern planes it will be forward of the Critical Point. Thus, the downward hydrodynamic force on the hull will be *less* than the upward force applied at the stern planes, and the submarine will *rise,* as shown in Fig. 3.36.

In order for the submarine to dive when it is travelling at such a very slow forward speed it is necessary to apply a *downward* force at the stern planes, and a downward force at the bow planes. This reversal in the effect of the stern planes is

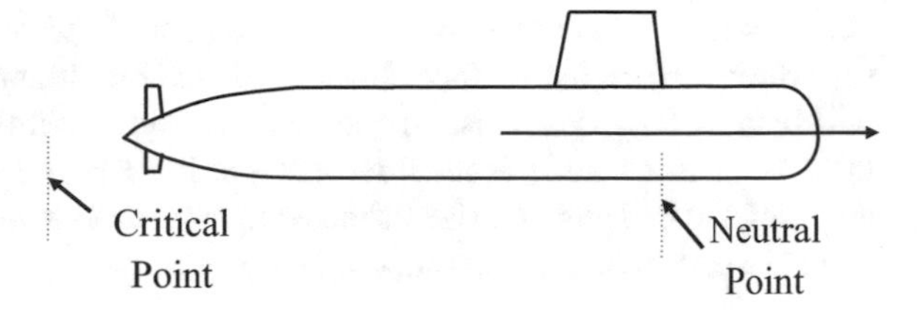

Fig. 3.35 Positions of neutral point and critical point for a submarine at very low speed

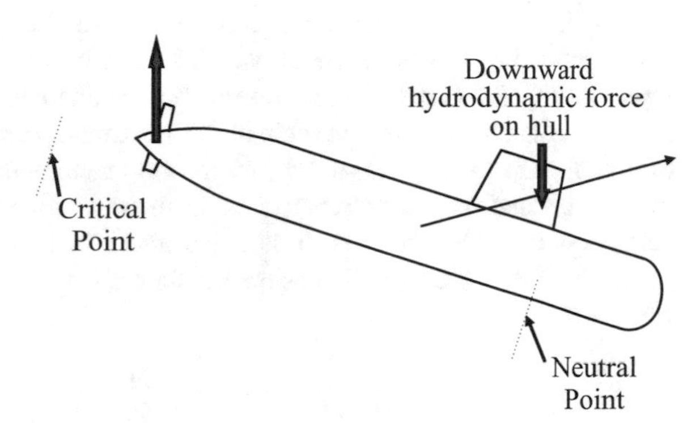

Fig. 3.36 Submarine attempting to dive using stern planes only when upward force at stern plane is forward of the critical point—submarine will rise!

counterintuitive, and can be disconcerting for those operating a submarine at these extreme low speeds.

Note that in addition, the position of the Neutral Point may well move aft at very low speeds, depending on the magnitude of the heave velocity compared to the forward velocity, as discussed in Sect. 3.7.3. This will increase the lever between the bow planes and the Neutral Point, increasing the ability of the bow planes to change trim angle.

Thus, the bow planes are very important for control of the submarine at low speeds.

3.8 Manoeuvring Close to the Surface

3.8.1 Surface Suction

When a submarine is travelling close to the water surface it will experience an upwards force, known as surface suction. This can have a significant effect on the behaviour of the submarine operating at periscope depth, and needs to be incorporated into any simulation of submarine motions. It also should be understood by operators as it can affect the safety of the submarine.

Surface suction can occur in calm water, but can also be greater in the presence of wind generated waves. When a submarine experiences surface suction it may be necessary to take on additional ballast to prevent it from broaching through the water surface. This must be done with care, as if the submarine moves away from the water surface the surface suction will reduce dramatically, and the submarine will then be "heavy". This, coupled with compressibility (discussed in Sect. 2.2.1) may result in an uncontrolled descent, particularly at low speeds when the planes are ineffective. Thus, the size of the control surfaces may be dictated by the need to control depth when at periscope depth in waves.

(a) Surface Suction in Calm Water

In calm water, the surface suction is due to the reduced volume above the submarine compared to below it, which results in a higher flow velocity, as illustrated in Fig. 3.37 (streamlines closer together) and hence a lower pressure. This is similar to the case of a surface ship in the horizontal plane when it is travelling close to a bank, or close to the seabed, where the phenomenon is known as squat.

Surface suction can be represented in the equations of the forces and moments acting on the submarine, such as those given in Eqs. 3.9–3.12 by making Z'_* a function of H^*, the non-dimensional distance from the water surface, defined in Eq. 3.136.

$$H^* = \frac{H}{D} \tag{3.136}$$

H is the distance from the water surface to the hull centreline, and D is the diameter of the submarine, as shown in Fig. 3.38.

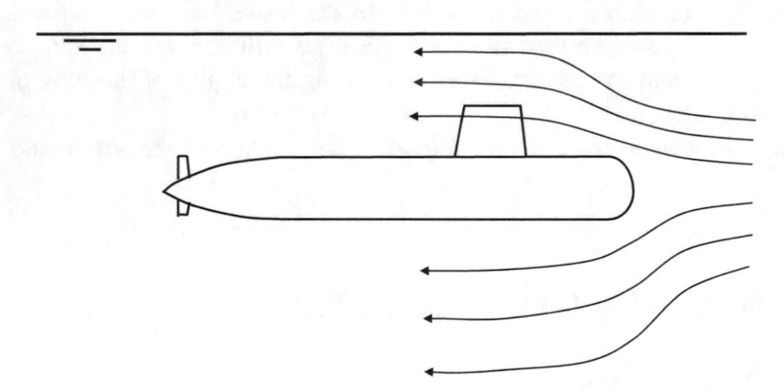

Fig. 3.37 Schematic of streamlines around a submarine travelling close to the surface

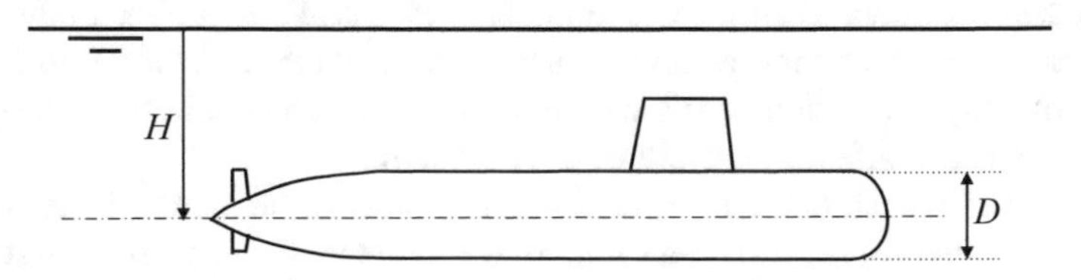

Fig. 3.38 Definition of H and D

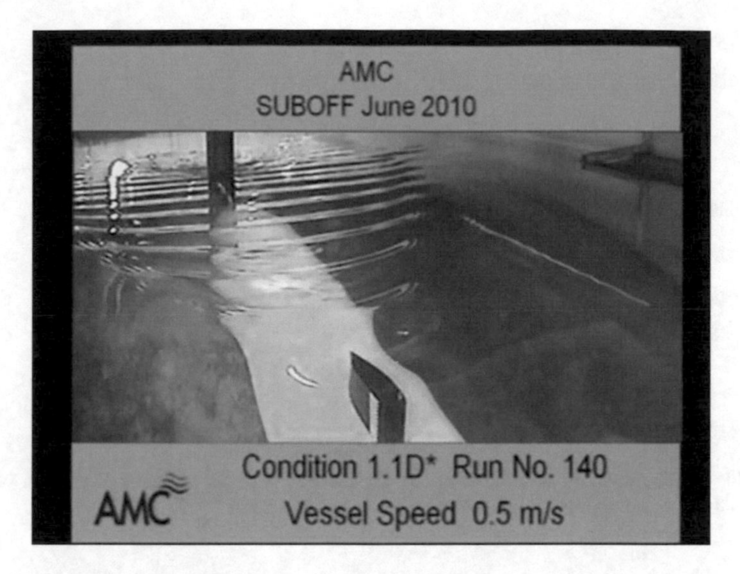

Fig. 3.39 Photograph of surface waves generated by a submarine model close to the surface. Model is towed using a sting, as shown in Fig. 3.4. $H^* = 1.1$; $F_r = 0.133$ (courtesy of the Australian Maritime College)

The additional complication compared to a surface ship travelling close to a boundary is that the free surface boundary is not flat. When the submarine is travelling close to the water surface, surface waves are generated, as shown in Fig. 3.39. This is discussed further in Chap. 4 in the context of the additional drag generated by a submarine close to the surface.

The wave pattern generated by the submarine complicates the surface suction effect. As these surface waves are a function of Froude number, F_r, the surface suction can be represented as given in Eq. 3.137.

$$\text{Surface suction} = Z'_*(H^*, F_r) \tag{3.137}$$

There will also be a pitch moment caused by proximity to the surface, and this can be represented by Eq. 3.138.

$$\text{Pitch moment} = M'_*(H^*, F_r) \tag{3.138}$$

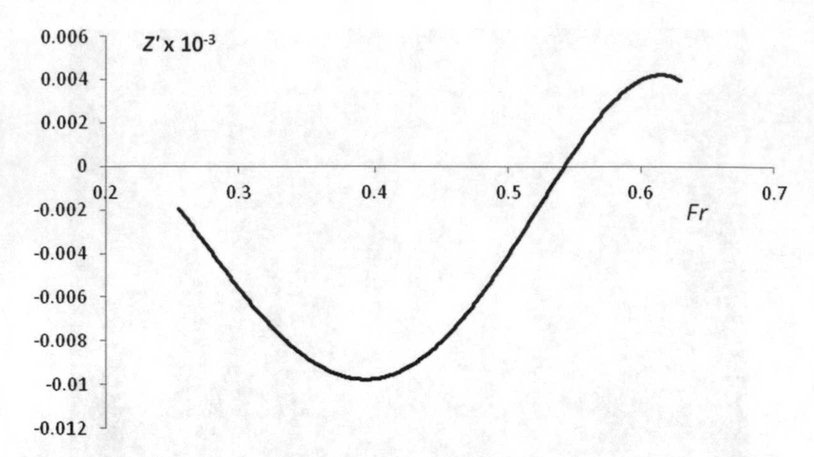

Fig. 3.40 Surface suction as a function of Froude number. Suboff geometry with sail only; $H^* = 1.5$ (taken from computational results presented in Griffin 2002)

An example of the surface suction as a function of Froude number for an H^* value of 1.5 is given in Fig. 3.40, taken from computational results presented in Griffin (2002) for the Suboff geometry (Roddy 1990) with sail but no other appendages.

Note that the suction force is upward (negative values of Z') for low Froude numbers, but that it becomes downward at higher Froude numbers. This is due to the complex wave pattern forming on the water surface above the submarine.

Results for the Suboff geometry without the sail for the heave force coefficient and the pitch moment coefficient are given in Fig. 3.41, taken from Renilson et al. (2014).

The longitudinal position of the centre of pressure for this case is given in Fig. 3.42. As can be seen the centre of pressure is always aft of amidships (negative value). In extreme cases it is even aft of the stern of the boat (non-dimensional position of −0.5). Thus, the planes are required to balance the boat, as shown in Fig. 3.43, as it is not possible to use additional ballast to do this.

When the upward force acts aft of the stern plane the bow plane will be required to generate an upward force to provide the required pitch moment. An example of the plane angles required to do this is given in Fig. 3.44, taken from Renilson et al. (2014).

(b) Surface Suction in the Presence of Wind Generated Waves

When a submarine is operating close to the water surface in the presence of wind generated waves then there will be oscillatory forces and moments at the wave frequency, as with a surface ship.

In addition, it will experience a wave induced low frequency (second order) surface suction effect which will depend on: speed; depth; sea state; and heading to the waves. Veillon et al. (1996) stated that for a 10,000 tonne submarine at a depth

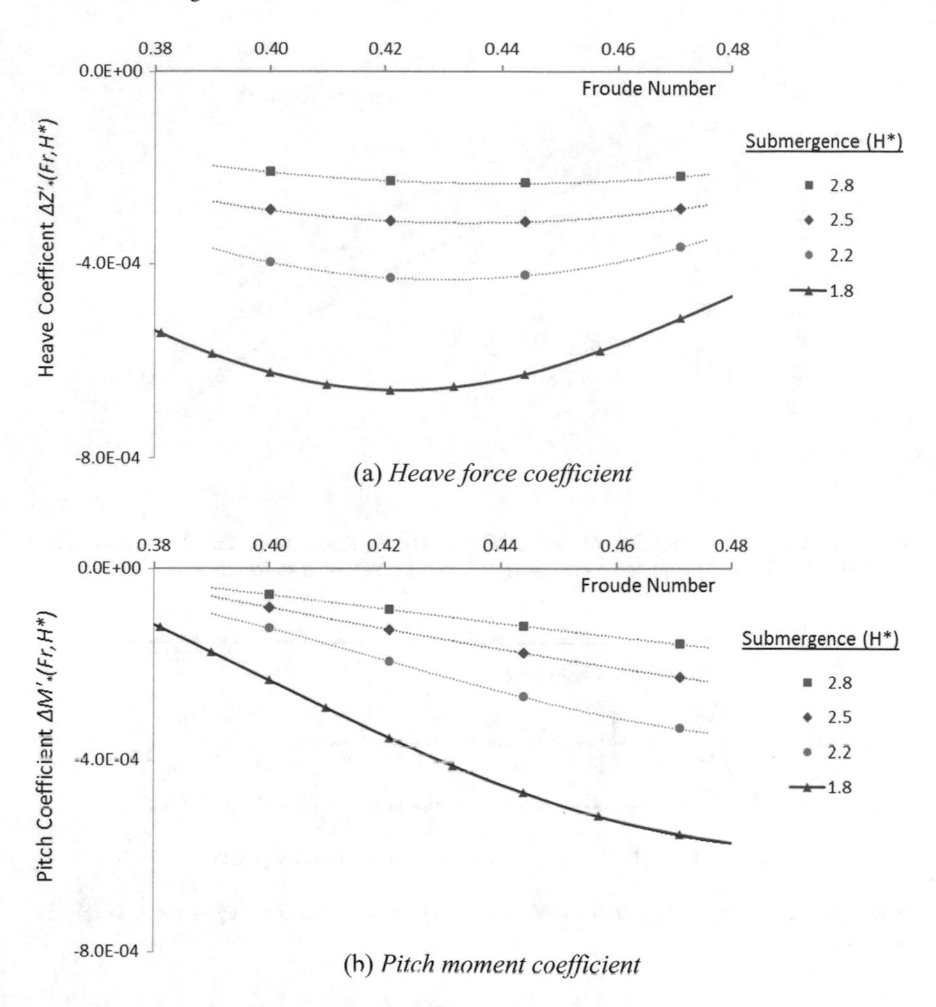

Fig. 3.41 Heave force and pitch moment coefficients as functions of Froude number suboff geometry with no sail (taken from computational results presented in Renilson et al. 2014)

of 50 m the surface suction due to waves will require around 20–30 tonnes of compensation to stop the submarine from surfacing.

Calculation of both the first order and the second order wave forces and moments can be carried out using potential flow, as discussed by Musker (1984).

The first order force and moment, at the same frequency as the waves, are functions of wave height, but the second order force and moment, are functions of wave height squared. Thus, they can be non-dimensionalised as given in Eqs. 3.139–3.142 (Crossland 2013).

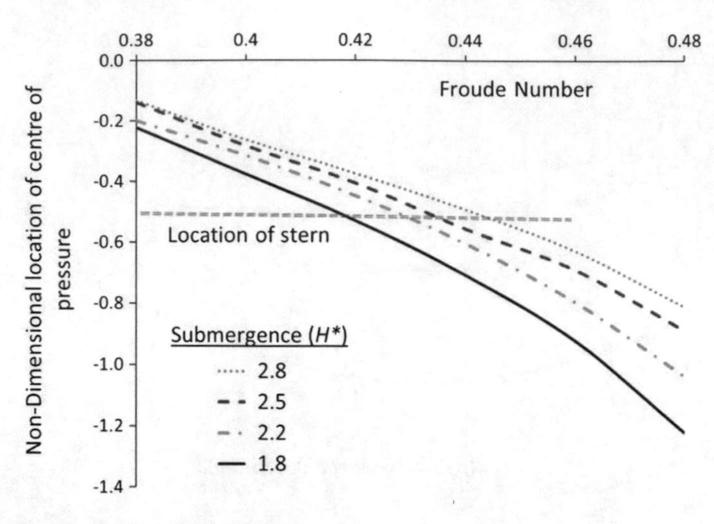

Fig. 3.42 Longitudinal position of centre of pressure as functions of Froude number suboff geometry with no sail (taken from computational results presented in Renilson et al. 2014)

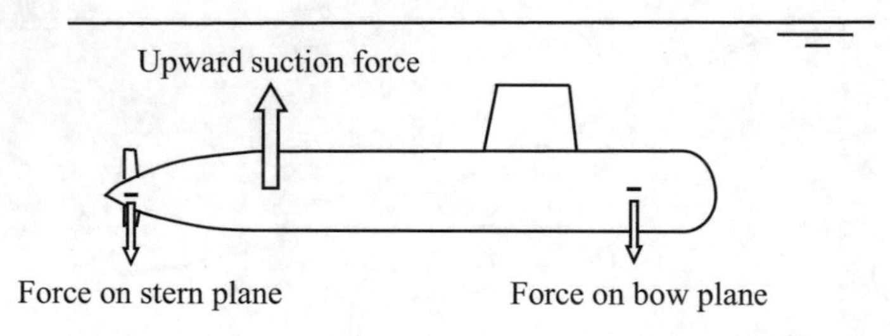

Fig. 3.43 Schematic of the additional forces acting on the submarine when close to the free surface

$$Z'_{RAO} = \frac{Z_{RAO}}{\rho g L^2 \zeta_w} \tag{3.139}$$

$$M'_{RAO} = \frac{M_{RAO}}{\rho g L^2 D \zeta_w} \tag{3.140}$$

$$Z'_{MEAN} = \frac{Z_{MEAN}}{\rho g L \zeta_w^2} \tag{3.141}$$

$$M'_{MEAN} = \frac{M_{MEAN}}{\rho g L D \zeta_w^2} \tag{3.142}$$

In Eqs. 3.139–3.142 Z_{RAO} is the heave force at the wave amplitude, M_{RAO} is the pitch moment at the wave amplitude, Z_{MEAN} is the second order heave force, M_{MEAN}

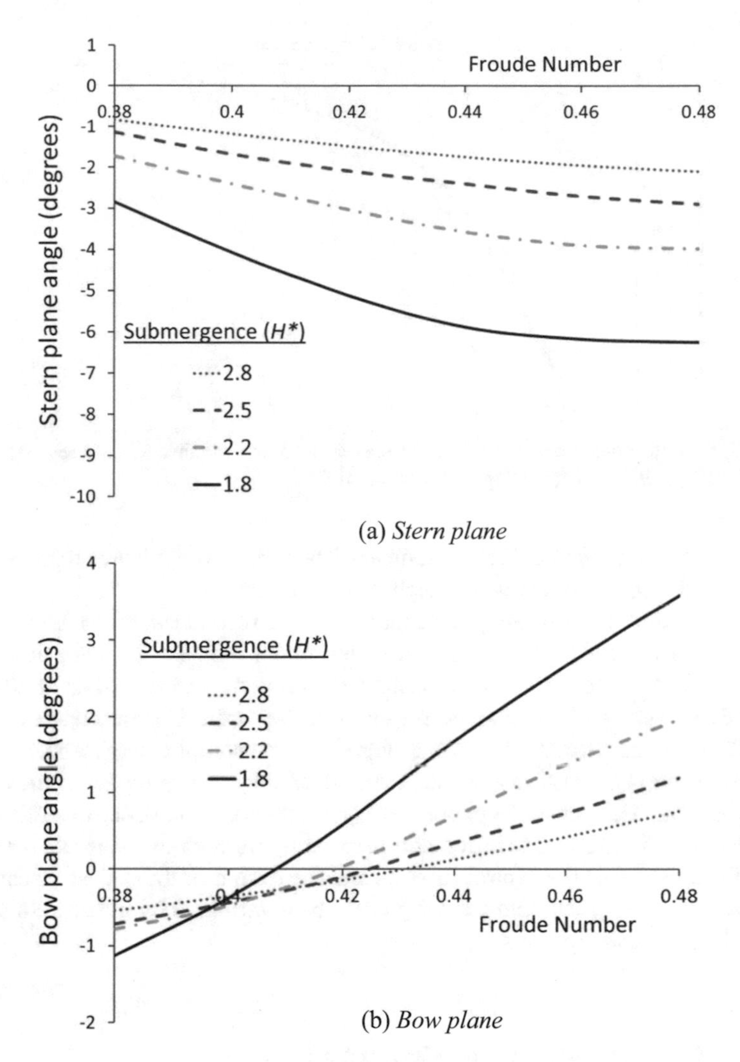

Fig. 3.44 Variation in plane angles required as functions of Froude number suboff geometry with no sail (taken from computational results presented in Renilson et al. 2014)

is the second order pitch moment, ρ is the water density, L is the submarine length, D is a representative diameter of the submarine, and ζ_w is the wave height.

Note that, due to non-linearities, the second order force and moment measured on a fully captive model is different to that measured on a model free to heave and pitch (Crossland 2013). The second order heave force as a function of (depth to keel)/ diameter is given in Fig. 3.45, adapted from Crossland (2013). This is for head seas at a Froude number of 0.08. At higher Froude numbers there will be an additional effect due to the calm water component, whereas at a Froude number of 0.08 the calm water component is small. This is because the calm water component is roughly a function of

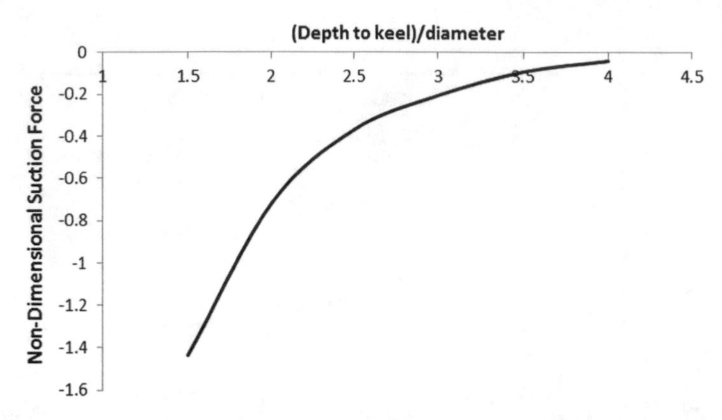

Fig. 3.45 Second order heave force in head seas at a Froude number of 0.08 sea state 5: $H_{1/3}$ = 3.87 m; T_0 = 9.7 s (adapted from Crossland 2013)

speed squared for moderate Froude numbers. The second order heave force is only a function of the square of the wave height (Crossland 2013).

There is a significant heading dependency on the second order heave force, with greater forces in head and following seas, but much reduced forces in beam seas, which may even be positive. This means a downwards force (Crossland 2013).

The dynamics of the submarine response in the vertical plane means that it is susceptible to the effects of wave grouping. For example, the submarine may encounter groups of large waves for a period of time, resulting in an increase in surface suction. This must be opposed by a combination of the control surfaces and the ballast system to avoid the submarine broaching through the water surface. This means that the size of the control surfaces and the design of the ballast system need to be considered together, and the design may be determined by the need to control depth when at periscope depth in waves.

3.8.2 *Manoeuvring in the Vertical Plane*

When a submarine is deeply submerged the forces and moments on it are dependent only on the motions (velocities and accelerations), propeller rpm, and appendage angles as discussed in Sect. 3.3. This was used to develop Eqs. 3.7–3.13. Thus, the hydrodynamic forces and moments are independent of the depth of the submarine, and the trim angle.

However, when a submarine is operating close to the surface the hydrodynamic forces and moments on it will be functions of the depth of the submarine, *H,* and the trim angle, τ, as defined in Fig. 3.46.

Note that the trim angle is not the same as the pitch angle, θ, since it is measured relative to the water surface, not the direction that the submarine is travelling.

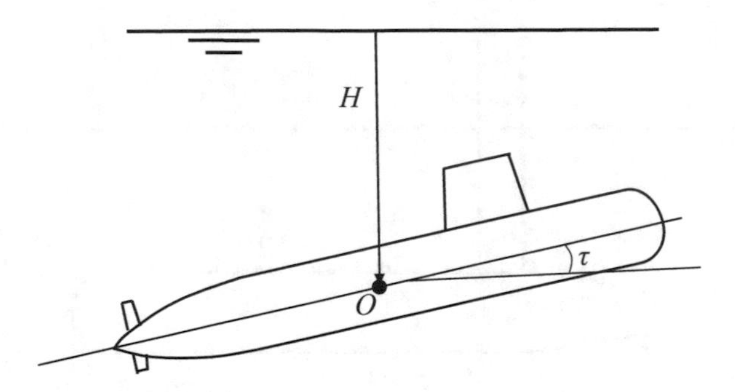

Fig. 3.46 Definition of trim angle, τ

Table 3.8 Additional coefficients required to represent the effect of the water surface on the hydrodynamic heave force and pitch moment

Effect	Coefficient	Equations	Comments
Heave force as a function of distance from water surface	$Z'_*(H^*, \tau, F_r)$	3.9	This coefficient already exists in Eq. 3.9 due to asymmetry, however when operating close to the surface it will be a function of H^*, τ, and F_r as discussed in Sect. 3.8.1.
Pitch moment as a function of distance from water surface	$M'_*(H^*, \tau, F_r)$	3.11	This coefficient already exists in Eq. 3.11 due to asymmetry, however when operating close to the surface it will be a function of H^*, τ, and F_r as discussed in Sect. 3.8.1.
Heave force as a function of trim angle	$Z'_\tau(H^*, \tau, F_r)$	3.9	This is a new coefficient. When operating deeply submerged, trim angle does not influence the hydrodynamic force, however, when close to the surface this effect needs to be taken into account.
Pitch moment as a function of trim angle	$M'_\tau(H^*, \tau, F_r)$	3.11	This is a new coefficient. When operating deeply submerged, trim angle does not influence the hydrodynamic moment, however, when close to the surface this effect needs to be taken into account.

However, in the special case where the submarine is moving parallel to the water surface, these are the same.

Thus, additional terms are required in Eqs. 3.9 and 3.11 to represent the effect of the presence of the water surface on the hydrodynamic forces and moments, as given in Table 3.8.

In addition, in principle when a submarine is operating close to the surface all the hydrodynamic coefficients in Eqs. 3.9 and 3.11, (which are for the vertical force, Z, and the pitch moment, M, respectively) should be functions of H^*, τ, and F_r. In

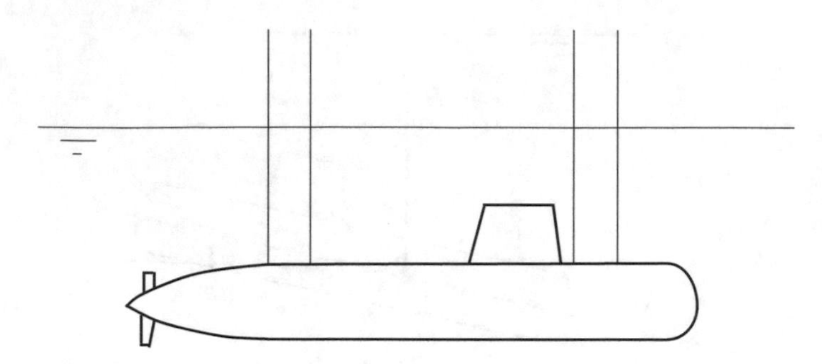

Fig. 3.47 Schematic of submarine being tested close to the surface using the conventional two strut support system

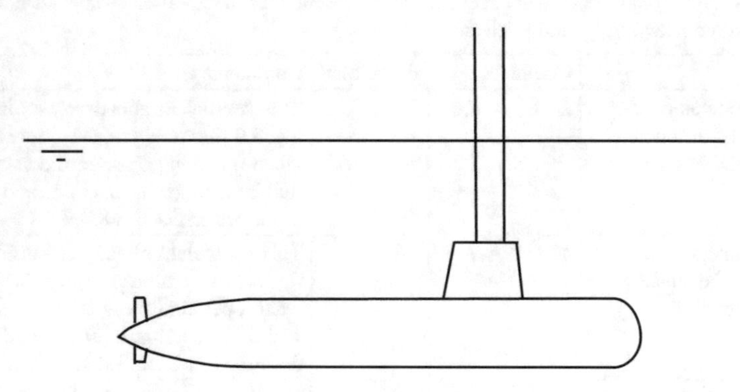

Fig. 3.48 Schematic of submarine being tested close to the surface using a single strut support system

practice it is not necessary to consider all of them as functions of H^*, τ, and F_r, as the effect of the relatively minor changes in some of these coefficients as the vessel is closer to the surface are negligible (Broglia et al. 2007; Polis et al. 2013).

Great care needs to be taken when conducting captive experiments to investigate the hydrodynamic forces and moments on a submarine model operating in the vertical plane close to the surface. Unlike the deep water situation, where the model can be tested inverted to avoid interference between the support struts and the sail (Fig. 3.3), it is necessary to test the model upright, resulting in a potential interference, as shown in Fig. 3.47. The effect of the interference can be investigated using CFD, and corrections made to the experimental results. Alternatively, a sting type arrangement, as shown in Fig. 3.4, can be utilised, however this does not permit an operating propulsor.

Another alternative is to use a single strut attached to the sail, as shown in Fig. 3.48.

3.8.3 Manoeuvring in the Horizontal Plane

As with manoeuvring in the vertical plane, discussed in Sect. 3.8.2, the existing coefficients in the equations used to obtain the hydrodynamic sway force (Eq. 3.8), roll moment (Eq. 3.10), and yaw moment (Eq. 3.12) should, in principle, all be functions of H^*, τ, and F_r. However, as with the vertical plane, the effect of the relatively minor changes in some of these coefficients as functions of distance from the water surface are negligible, and do not need to be taken into account for typical manoeuvres.

Additional terms which are required in the equations representing the hydrodynamic forces and moments in the vertical plane, are not required in the horizontal plane, other than those to represent the wave resistance due to the proximity to the water surface, as discussed in Chap. 4.

Conducting captive model experiments to investigate the hydrodynamic forces and moments when operating in the horizontal plane close to the water surface is difficult. It is not possible to test the model on its side, as done for deep water. As the model needs to be tested at a range of drift angle in a towing tank, the conventional two strut support system would require the struts to be aligned with the flow for each drift angle. Equally, care needs to be taken when testing close to the surface using a rotating arm to avoid interaction with the support struts.

An alternative is to use a Horizontal PMM (HPMM) with the model attached using a sting, as shown in Fig. 3.49. Here, the model origin is located directly below the PMM support strut, and the sting curved to attach to the model at its stern. With this mechanism it is not possible to have a rotating propulsor, however it may be adequate to investigate the influence of the water surface on the coefficients representing the hydrodynamic force and moments on the submarine in the horizontal plane.

3.9 Manoeuvring Criteria

A recommended range for the various manoeuvring criteria is given in Table 3.9, taken from Ray et al. (2008). This is adequate at the initial design stage to determine the required size of both fixed and control appendages.

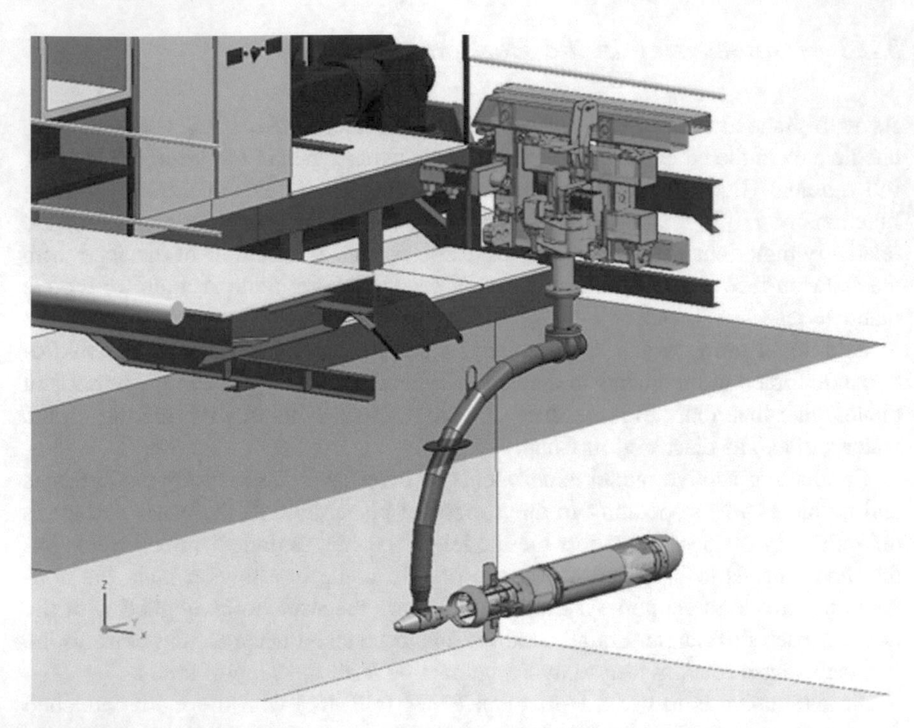

Fig. 3.49 Horizontal PMM set up for captive model tests using a sting support (courtesy of Australian Maritime College)

Table 3.9 Recommended range of stability and control indices (taken from Ray et al. 2008)

Parameter	Expression	Acceptable range
Vertical stability index	G_V	0.5–0.8
Horizontal stability index	G_H	0.2–0.4
Stern planes heave effectiveness	$Z'_{\delta S}/\left[(0.001L)\left(Z'_{\dot{w}} - m'\right)\right]$	2.5–4.5
Stern planes pitch effectiveness	$M'_{\delta S}/\left[(0.001L^2)\left(M'_{\dot{q}} - I'_{yy}\right)\right]$	0.2–0.4
Bow planes heave effectiveness	$Z'_{\delta B}/\left[(0.001L)\left(Z'_{\dot{w}} - m'\right)\right]$	0.7–1.7
Bow planes pitch effectiveness	$M'_{\delta B}/\left[(0.001L^2)\left(M'_{\dot{q}} - I'_{yy}\right)\right]$	−0.8 to −0.2
Rudder sway effectiveness	$Y'_{\delta R}/\left[(0.001L)\left(m' - Y'_{\dot{v}}\right)\right]$	3.0–5.0
Rudder yaw effectiveness	$N'_{\delta R}/\left[(0.001L^2)\left(N'_{\dot{r}} - I'_{zz}\right)\right]$	0.2–0.6

3.10 Manoeuvring Limitations

3.10.1 Introduction

The safe operation of submarines requires a comprehensive understanding of submarine dynamics, in particular the dynamics associated with emergency manoeuvres, and extreme operations.

Manoeuvring limitations are placed on a submarine to ensure that it is operated in such a way that it can survive a credible failure. Safe Operating Envelopes (SOEs) are generated, and used by the crew to ensure that the submarine does not get into a situation from which it cannot recover if a credible failure occurs.

Typical credible failures include the following:

(a) permanent plane jams, usually at the last ordered angle;
(b) temporary plane jams at the maximum angle;
(c) runaway plane, where the plane moves to the hard-over position due to a failure of the control system;
(d) securable and unsecurable floods; and
(e) failure of power.

As part of the submarine safety assessment, these, and any other credible failures, must be determined, and the probability of occurrence obtained.

In addition, there is a need to investigate critical operations, such as when at low speed close to the surface, and apply constraints to the operation of the submarine as appropriate.

To generate an SOE, it is necessary to have an accurate numerical model of the manoeuvring of the submarine. This must be capable of simulating a large number of manoeuvres in a very short time. As this is a safety critical element, such software must be correctly verified, validated, and accredited.

In particular, it is vital that the numerical model is capable of simulating extreme manoeuvres close to the boundary of the SOE with a high degree of accuracy.

3.10.2 Safe Operating Envelopes

SOEs are restrictions placed on the operation of the submarine to ensure that it does not get into a situation from which it cannot recover if a credible failure occurs.

These limitations can be depicted using either Manoeuvring Limitation Diagrams (MLDs), or Safe Manoeuvring Envelopes (SMEs), as discussed by Marchant and Kimber (2014).

MLDs consist of plane limitations, where a range of plane angle limits are displayed on a single diagram for a given initial pitch angle, as shown in Fig. 3.50. Further diagrams may be used for different initial pitch angles, and/or different initial trim conditions.

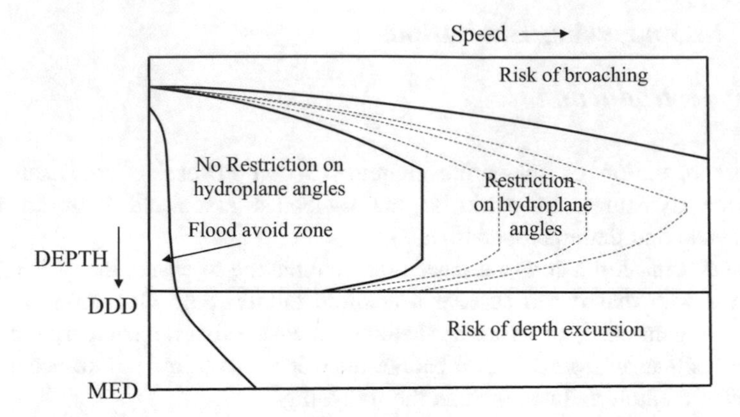

Fig. 3.50 Example of manoeuvring limitation diagram

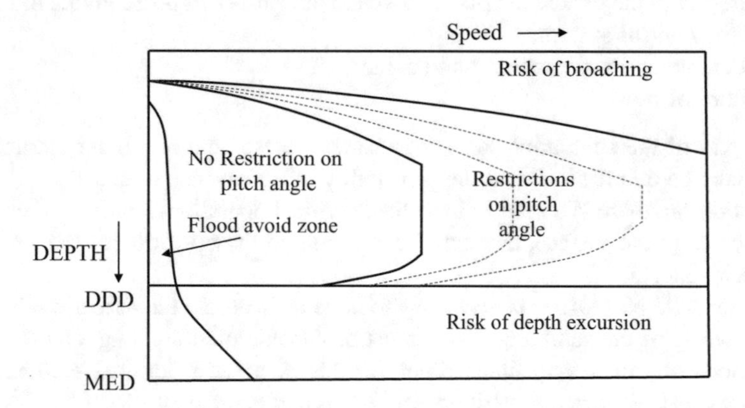

Fig. 3.51 Example of safe manoeuvring envelope

SMEs consist of pitch limitations, where a range of limiting pitch angles are displayed on a single diagram for an assumed jam angle as shown in Fig. 3.51. Further diagrams may be used for different jam assumptions and/or different trim conditions.

3.10.3 Generation of Safe Operating Envelopes

To generate an SOE, it is necessary to have a manoeuvring simulation model of the submarine. It is also necessary to know the Standard Operating Procedures (SOPs) which the crew will conduct in the event of each failure, including an estimate of the crew reaction time. Different realistic SOPs can be investigated, and advice on the appropriate SOP for particular incidents can be developed, in conjunction with the operators.

SOEs are then generated by running the simulations for a wide range of conditions and assessing the trajectory of the submarine to determine whether or not it will survive. This needs to be repeated a very large number of times to generate all the boundaries shown in Figs. 3.50 and 3.51.

For this reason an accredited manoeuvring simulation model is required which can be run very quickly, as there are numerous conditions to assess.

A factor of safety must be applied to the results to account for unknowns such as: the inaccuracy of the manoeuvring simulation model; the crew reaction time; the submarine trim and ballast condition; and the initial conditions.

3.10.4 Aft Plane Jam

A typical SOP for an aft plane jam on a submarine with a cruciform stern is as follows:

(a) order full astern rpm;
(b) operate bow planes if they are available; and
(c) hold the rudder amidships.

Thus, to be able to simulate the resulting motions of the submarine after a stern plane jam it is necessary to understand the behaviour of the prime mover (how long it will take to generate power astern, and the relationship between power and rpm), and the hydrodynamic forces on the propulsor when going astern.

Failure after a stern plane jam will occur if the submarine:

(a) exceeds Maximum Excursion Depth (MED); or
(b) hits the seabed; or
(c) reaches the surface proximity limit; or
(d) develops an excessive trim angle.

Success after a plane jam incident is defined as:

(a) MED/seabed or proximity to the water surface not exceeded; and
(b) maximum permissible trim angle is not exceeded; and
(c) a trajectory in the opposite direction to the jam is achieved; and
(d) post recovery speed is maintained.

3.10.5 Flooding

Flooding can occur through systems connected to the sea which are normally kept closed.

A typical SOP for a flooding incident is as follows:

(a) secure the submarine, if possible;
(b) apply full forward rpm, if available;
(c) blow all main ballast tanks; and
(d) drive the submarine to the surface.

For diesel electric submarines flooding may result in the loss of power if the flood occurs in a compartment containing propulsion equipment, including batteries. This needs to be taken into account in the simulation.

Success after a flooding incident is defined as:

(a) MED or seabed not exceeded; and
(b) maximum permissible trim angle is not exceeded; and
(c) submarine must reach the surface with an acceptable level of reserve buoyancy.

High rates of rise, and unpowered ascents where the submarine may ascend stern first, are difficult to model, and great care needs to be taken when a simulation indicates that this may occur. According to Watt and Bohlmann (2004) during a buoyant ascent horizontal plane stability is compromised by:

(a) high incidence angle, especially at high speed;
(b) both positive and negative q and $\dot{q}$ values; and
(c) blowing only the forward ballast tanks.

What this means is that emergency rises are best carried out using a high pitch angle to increase speed, thereby reducing incidence angle. This will delay the onset of instability, and reduce the time that the instability has to act. Combining high speed and pitch changes increases the likelihood of instability, so once the desired pitch angle is achieved, it is recommended to keep this fixed until the boat is surfaced, and not to level off prior to surfacing. If it is necessary to blow the forward tanks first, blow the remaining tanks as soon afterwards as possible (Watt and Bohlmann 2004).

3.10.6 Operating Constraints

The manoeuvring limitations imposed by the SOEs place operational constraints on the submarine, hence the importance of ensuring that they are calculated accurately, with minimum conservatism due to lack of knowledge of the manoeuvring characteristics of a submarine in extreme conditions. Consequently, a lot of work has gone into understanding, and predicting, the behaviour of submarines in emergency situations.

The magnitude of the limitations placed on the operation of a submarine need to be understood at the design stage, such that its operational effectiveness can be assessed correctly.

3.11 Free Running Model Experiments

Free running model experiments can be used to investigate the manoeuvring characteristics of a submarine at a comparatively low cost in a controlled environment. The purposes of such experiments can be to:

(a) assess the manoeuvring characteristics of an existing or proposed design, including the effect of possible changes to the design, or in-service damage, including total or partial control surface failure;
(b) investigate different control strategies;
(c) explore behaviour in extreme situations, including emergencies, and validate emergency procedures and Safe Operating Envelopes;
(d) provide data for validation of a numerical model; and
(e) provide data for application of systems identification techniques.

A number of nations which operate submarines have the capability to conduct free running model experiments including: the British (Crossland et al. 2014); the Dutch (Overpelt 2014); the French, (Itard 1999); and the US (Fox 2001). The ability to operate large free running models is seen as an essential tool in ensuring submarine operational safety.

Typically free running submarine models are around 5 m long. Smaller models are considered to suffer too much from scale effects, as their control surfaces would be very small. Even with 5 m long models care has to be taken with scale effects, as the control surfaces have low Reynolds numbers and operate in a much larger (scaled) boundary layer.

Normally free running submarine models need to be at least semi-autonomous, as it is often not possible to send control commands to them. They can be designed as a compartmentalized aluminum pressure hull containing all the instrumentation with GRP cladding that can be modified to represent the required hull shape and control surface configuration (Haynes et al. 2002; Crossland et al. 2011). A low positive pressure can be maintained in the pressure hull to reduce the consequences of a small leak. A schematic of such a free running model is given in Fig. 3.52.

An alternative approach to the single aluminum pressure hull is to make use of a space frame construction with individual components located in their own pressure resistant compartments. This has been developed recently by QinetiQ in the UK, making it easier to replicate submarines with large *L/D* ratios, such as SSBNs.

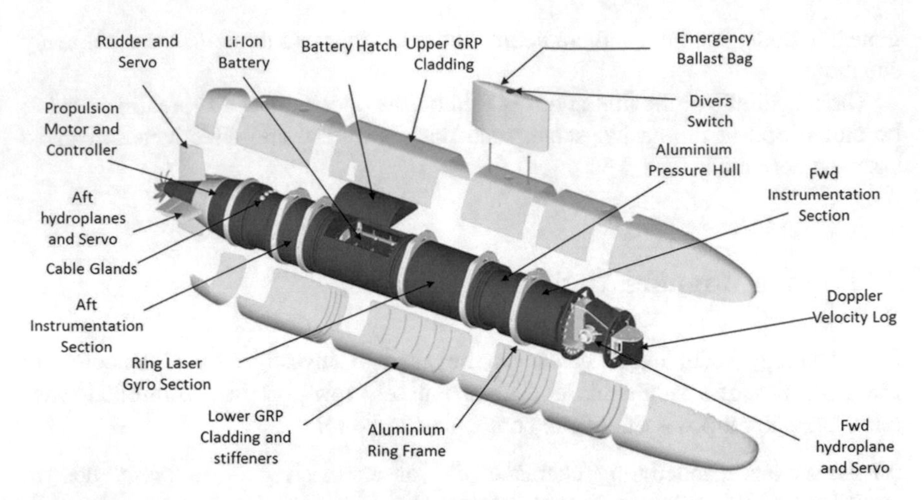

Fig. 3.52 Schematic of typical free running submarine model with single pressure compartment (courtesy QinetiQ Limited, © Copyright QinetiQ Limited 2017)

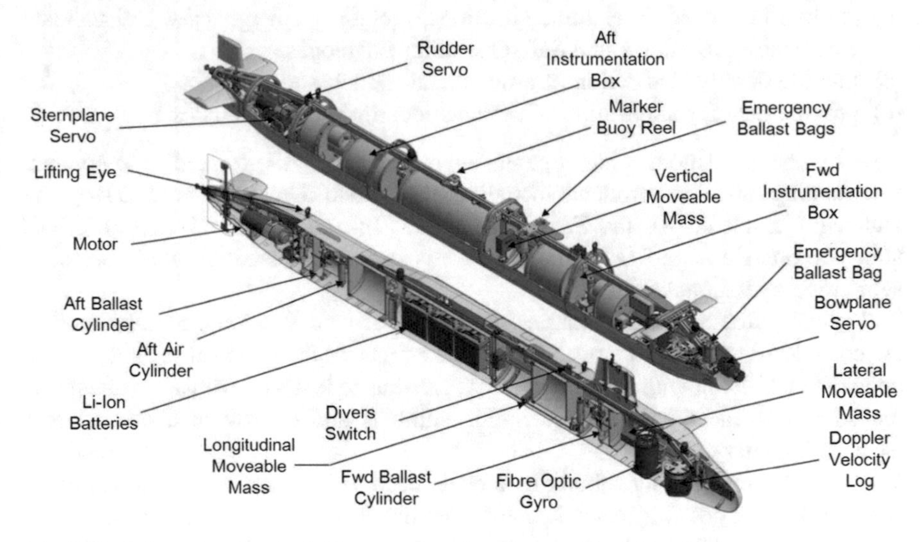

Fig. 3.53 Schematic of typical free running submarine model with multiple pressure compartments (courtesy QinetiQ Limited, © Copyright QinetiQ Limited 2017)

(Crossland et al. 2014). A schematic of such a free running model is given in Fig. 3.53.

Free running models are usually fitted with the following equipment:

(a) propulsion motor with controller and rpm measurement;
(b) aft planes, forward planes, and rudders, with servos, controllers and deflection angle measurement;

(c) trim and ballast system (movable mass and ballast systems);
(d) inertial position measurement system;
(e) doppler velocity log;
(f) depth/pressure measurement;
(g) autopilot;
(h) data logging and recording equipment;
(i) battery power pack;
(j) model/shore connection to permit downloading of data, loading of the next exercise, and charging of the battery;
(k) emergency recovery ballast system; and
(l) emergency locator system.

The model must have a mechanism to permit fine adjustments to its mass/buoyancy balance, both in the vertical and longitudinal planes, as the ability of the model to be in balanced trim is essential. Also, the ability to vary the vertical centre of gravity, such that a realistic BG can be replicated may be important. It is also important to know its inertial properties. Figure 3.54 shows a model being prepared using a compound pendulum to measure its inertia.

Fig. 3.54 Model being prepared using a compound pendulum to measure inertia (courtesy QinetiQ Limited, © Copyright QinetiQ Limited 2017)

Fig. 3.55 Free running model being recovered by support boat in lake testing facility (courtesy QinetiQ Limited, © Copyright QinetiQ Limited 2017)

If compressibility considerations are to be investigated then a means of replicating the compressibility characteristics of the full scale boat needs to be considered. This may be able to be achieved using the trim and ballast control system.

The model should be fitted with a kill switch to permit divers, and others, to shut down the motor if required. An emergency ballast system and emergency locator system should be provided for operation in a lake, to enable recovery of the model in the event of a failure.

Finally, the model should be fitted with a mechanism to facilitate handling in and out of the water, including from the support boat when on the surface (Fig. 3.55), and a suitable cradle should be provided (Fig. 3.56).

Where possible, commercial off the shelf (COTS) equipment should be used and of course adequate spares should be available such that maintenance and repair of malfunctioning equipment can be carried out during the test period. The recent advances in such technology make this much easier than was possible only a few years ago, and potential future advances in this area should be considered in the design. Ideally the pressure hull must be able to be opened reasonably easily on site to allow access to the on board equipment.

Free running submarine models can either be operated in existing hydrodynamics test basins, Fig. 3.57, or in suitable lake test facilities, Fig. 3.58. The advantages of using hydrodynamics test basins are that the environment can be

Fig. 3.56 Free running model on purpose built cradle (courtesy QinetiQ Limited, © Copyright QinetiQ Limited 2017)

Fig. 3.57 Free running submarine model with cradle in hydrodynamic test basin (courtesy QinetiQ Limited, © Copyright QinetiQ Limited 2017)

Fig. 3.58 Free running submarine model in lake test facility (courtesy QinetiQ Limited, © Copyright QinetiQ Limited 2017)

controlled (in particular wave generation, for investigating the effect of waves on the submarine behaviour) and that all associated support staff and equipment are close to hand. In addition, staff members do not have to travel from their "home base". Further, it may be possible to set up an ability to communicate with the model, and it is generally visible during the tests.

On the other hand, few hydrodynamics test basins are deep enough to permit depth changing manoeuvres, and hence any serious free running model capability must have access to a suitable lake test facility.

The requirements of a lake test facility include the following:

(a) sufficient area to permit a range of manoeuvres without encroaching on the shallow areas at the edges: for a 5 m model an area of at least 200 m × 200 m of deep water is required, but greater area is desirable, in which case multiple manoeuvres can be carried out in each run (Fig. 3.58);
(b) suitable depth to permit depth changing manoeuvres, but not too deep to result in the loss of the model in the event of a failure: for a 5 m model a depth of 20–30 m is ideal;
(c) ability to easily launch and recover the submarine model and the support boat;
(d) suitable office and workshop accommodation, with power, for the trials crew, and secure storage for the model and equipment;

(e) facilities for calibrating measuring equipment, if necessary;
(f) facilities for diver operations, if required;
(g) adequate shelter, in particular from wind and wind generated waves;
(h) lack of water currents;
(i) temperature and density consistency; and
(j) sufficient privacy, preferably with no other users, when the free running trials are being conducted.

In addition the lake must be either sufficiently close to the "home base" to enable staff to return home each evening, or be located close to amenities such as hotel accommodation and restaurants/shops, including hardware outlets. It should also be sufficiently accessible to clients and other personnel who require brief visits during the test periods.

3.12 Submarine Manoeuvring Trials

3.12.1 Introduction

Full scale manoeuvring trials are conducted on submarines for the following different purposes:

(a) to confirm whether design specifications have been met (either on the initial design, or after modifications); and
(b) to generate data to improve future predictions of submarine manoeuvring.

The generation of data is particularly important, especially to help to understand the correlation between model and full scale manoeuvring performance.

3.12.2 Definitive Manoeuvres

Typical definitive manoeuvres used for submarine trials are given in Table 3.10. Note that prior to initiation of each trial the boat must be in balance, and travelling at a steady speed on a steady heading/depth for sufficient time to be confident that the initial condition is stable.

Also, for tests in the horizontal plane it may be necessary to operate the forward and aft planes to ensure that the boat remains at constant trim and depth, as the asymmetry due to the casing and the sail will result in a vertical force and pitching moment when turning, as discussed in Sect. 3.6.6.

Typical results from these definitive manoeuvres are given in Figs. 3.59, 3.60, 3.61 and 3.62.

Table 3.10 Typical definitive manoeuvres used for submarine trials

Trial	Description	Measured quantities	Purpose	Comments
Turning circle	The rudder is set at a constant angle and the boat executes a full 360° heading change.	Tactical diameter; advance; transfer; drift angle; heel angle; and speed loss (Fig. 3.59).	Steady turning ability.	The primary results are not very sensitive to initial speed.
Pull out	When the boat is in a steady turn (for example at the end of a turning circle) the rudder is returned to amidships and the residual yaw rate of the vessel is recorded.	Yaw rate as a function of time (Fig. 3.60).	To assess straight line stability.	Can be conducted at the completion of a turning circle.
Zig-zag/ overshoot	For a zig-zag in the horizontal plane the rudder is deflected to a constant angle, δ_0, as quickly and as smoothly as possible and maintained at this angle until the change in the boat's heading becomes ψ_O. The rudder is then deflected to $-\delta_O$ and held steady until the boat's heading has changed to $-\psi_O$. A typical value of δ_O and ψ_O is 20° in which case this is known as a 20–20 zig-zag. Other angles can be used instead. For a zig-zag in the vertical plane the stern planes are deflected instead of the rudder, and a specified change in pitch is used instead of change in heading. This is often referred to as an overshoot manoeuvre.	Time at which each change in control surface angle is initiated; yaw/pitch overshoot; width of path/depth overshoot; time to maximum width/ depth excursion (Fig. 3.61).	Indicates course-changing (horizontal plane) or depth-changing (vertical plane) ability.	Results are dependent on speed, as rudder/plane rate is constant.

(continued)

Table 3.10 (continued)

Trial	Description	Measured quantities	Purpose	Comments
Meander	This is similar to the zig-zag manoeuvre in the vertical plane, with the difference that once the execute pitch angle is reached, the control surfaces are brought to zero (and not reversed as in the zig-zag or overshoot test).	Path of vessel.	Indicates motion stability and depth keeping ability.	
Spiral	The rudder is deflected to a given maximum angle, such as 25°, and held until a steady yaw rate is obtained (A in Fig. 3.62). It is then reduced by a fixed increment (such as 5°) and held until a steady yaw rate is obtained. This is repeated until the opposite value of the initial maximum angle is reached (F in Fig. 3.62) with the steady yaw rate recorded at each increment. The process is then repeated in the opposite direction until the initial maximum rudder angle is achieved (A in Fig. 3.62).	Steady yaw rate for each rudder angle (Fig. 3.62).	Control authority and straight line stability. The size of the hysteresis loop, BDEH shown in Fig. 3.61b is a measure of instability.	Care needs to be taken to ensure that the yaw rate is steady at each increment. The results are not very sensitive to initial speed. This test can be very time consuming, and does not give information within the hysteresis loop for unstable boats.
Reverse spiral	The rudder angle required to achieve a given yaw rate is determined. This is then repeated for a range of different yaw angles.	Steady yaw rate for each rudder angle.	Control authority and straight line stability.	The results are not very sensitive to initial speed. For an unstable boat this gives information within the hysteresis loop.

3.12.3 Preparation for the Trials

Prior to conducting the trial it is very important to determine exactly the purpose of the trial. For example, is it purely to determine whether a new boat meets the design specification, or is it to generate new data to validate a manoeuvring code? If the former, it is clear which boat the trials need to be conducted on, whereas if the latter then the submarine selected will depend on considerations such as availability.

Once the purpose of the trial has been agreed, then trials orders need to be drawn up. These will include the following (Bayliss et al. 2005):

(a) scope and high level requirement;
(b) equipment fit required;
(c) trial programme;
(d) trials team and civilian personnel;
(e) detailed list of runs, including all requirements for each run; and
(f) risk assessment/safety case.

Determination and proper specification of the trials equipment is essential, including the required accuracy of any measuring instrumentation. It may be possible to make use of the existing submarine platform management system to measure, and record, some of the required data (Tickle et al. 2014). Whether this is possible, or not, and the effect that it could have on operations, needs to be understood at an early stage.

Also, any devices which affect the movement of the control surfaces need to be agreed upon well in advance, and taken into account in the risk assessment/safety case for the trial.

When planning the trials, and even when selecting which boat to use, consideration needs to be given to the boat′s operational program in the months prior to the trial. Specialist equipment may be required, and this will require the submarine to be available for such equipment to be fitted. It may also need to be removed immediately following the trial. The time and cost of doing this needs to be included in the plans.

It may also be necessary to conduct a number of “dummy runs” with the equipment on land prior to being fitted to the submarine. Sufficient spares should be carried on board, and at least one member of the trials team should be able to carry out diagnostic tasks and repairs of the equipment when on board the submarine, as otherwise a small malfunction can void the whole trial.

Careful thought needs to be given to how the results of the trial will be analysed, and hence the required accuracy of the measuring equipment, and the frequency of the analogue to digital conversion, if appropriate. In addition, a scheme for conducting a preliminary “running analysis” of the data on board the submarine should be developed. Such a running analysis is essential to ensure that all the channels are working correctly, and that the results are valid.

It may be advantageous to demonstrate the trials in a manoeuvring simulator in advance to the submarine’s senior officers to ensure that they fully understand what

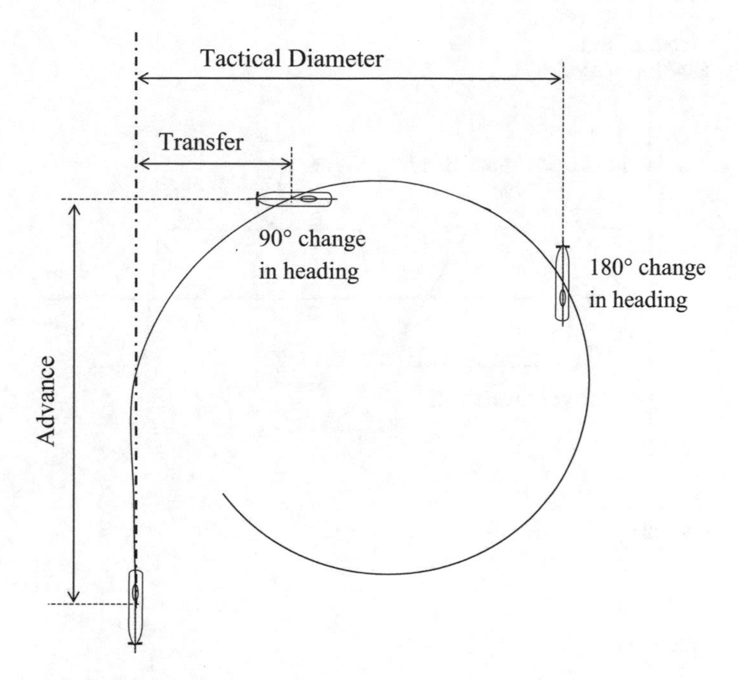

Fig. 3.59 Turning circle

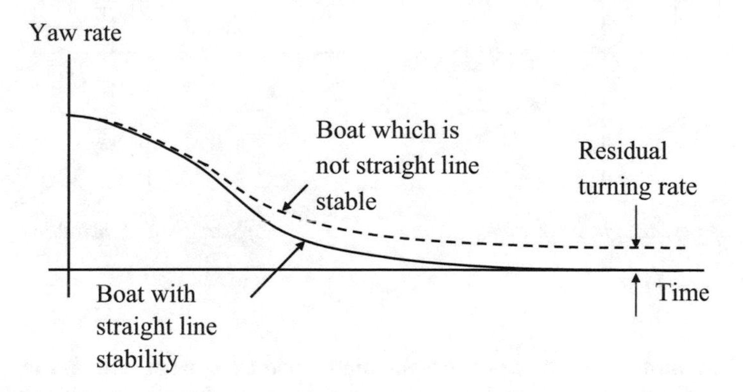

Fig. 3.60 Pull out

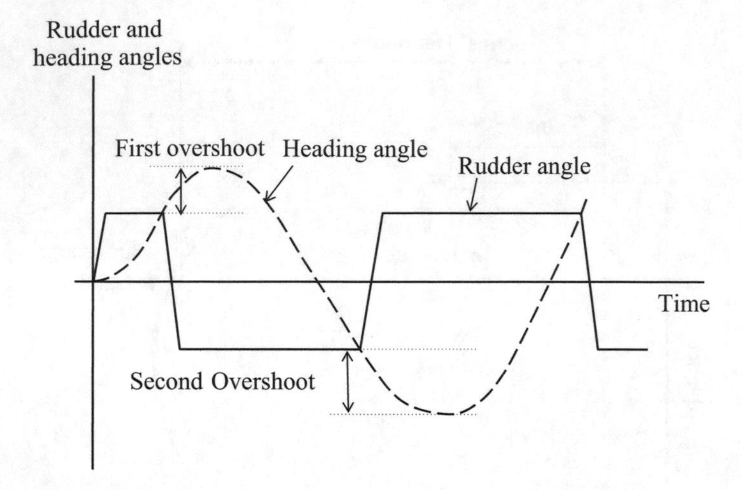

Fig. 3.61 Zig-zag

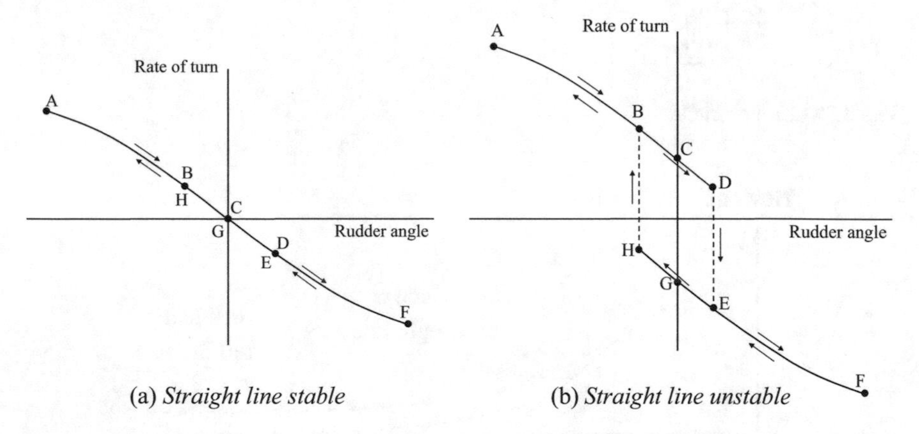

Fig. 3.62 Spiral test

is required, and to get advice at the planning stage as to what will and will not be permitted. Such a demonstration will also assist with those planning the trials to understand what can realistically be achieved in the available timescale.

A comprehensive risk assessment/safety case is required, and should be included as part of the initial development of the runs to be included since it may affect what can be achieved. This should include the following:

(a) the installation, diagnostic operations, and removal of the trials equipment (for example any confined space issues);
(b) the impact of the trials equipment on the safety of the boat (for example any influence of restraints to the movements of the controls);

(c) the safety of each of the runs, particularly where the boat is being taken close to the boundaries of any SOEs;
(d) the expected results from each run, and the need to abort a run, and the method of doing this, if the manoeuvre exceeds expected parameters; and
(e) the danger posed to civilian trials team members caused by submarine operations (for example it may be necessary for all trials team members to undergo training on submarine escape and rescue).

3.12.4 Conduct of the Trials

The harmonious interaction between the trials team and the submarine crew is vital to the successful completion of the trials. The members of the trials team are responsible for observing, and overseeing the trials, rather than to be carrying them out themselves. This is necessary, as the safety of the submarine is in the hands of the crew. In addition to the interaction during the planning phase the trials team will be living with the submarine crew in a confined space on board the submarine, so the development of a good working relationship between the trials team and the crew is essential.

Generally the trials will be conducted in stages at suitable opportunities alongside the operational tasks being undertaken by the boat. It is vital that each run is not rushed, and that proper initial conditions are established (and recorded) prior to the commencement of the run. The balance of the boat prior to each run must be known accurately, and unless specifically required, the boat should be as close to balance as possible. Also, it is essential that adequate time is taken prior to the execute command to ensure that the initial condition is steady. If this is rushed it is likely to completely invalidate the result from that run. The parameters prior to the execute command should be recorded, such that confidence in the initial conditions can be achieved when conducting the full analysis at a later date.

Where possible a preliminary "running analysis" should be undertaken on the results of each trial. Although this may only be fairly rudimentary it will ensure that all the data recording is functioning correctly and that the results are valid. It is very risky to proceed to the subsequent run before the data from the previous one has been subjected to this running analysis, as if one or more channel is not working then the complete run will be wasted. If time pressures are such that it is essential to do this then a quick check of the data must be made to ensure that all the channels are working, and the running analysis should be conducted as soon as practical after a group of runs. Backups need to be taken at suitable periods during the trials.

Table 3.11 gives an indication of the length of time likely to be required for various definitive trials manoeuvres.

Table 3.11 Typical test program used for submarine manoeuvring trials (adapted from Ray 2007)

Trial	Number of variants (for each speed)	Number of speeds	Approx total time
Overshoot (zig-zag) in vertical plane (dived)	One to five sets of plane angles for each speed, repeated using bow/stern planes alone and in combination and for both rise/dive	5	8½ h dived
Meander test in vertical plane (dived)	One; repeated using bow/stern planes alone and in combination and for both rise/dive	2	
Overshoot (zig-zag) in horizontal plane (dived)	Four sets of rudder angles for each speed; repeated for port as well as starboard deflection	5	8½ h dived
Turning circle (dived)	One to two rudder angles for each speed; repeated for port as well as starboard deflection	4	
Spiral manoeuvre in horizontal plane (dived)	One	3	
Overshoot (zig-zag) manoeuvre in horizontal plane (surfaced)	Two to three sets of rudder angles for each speed; repeated for port as well as starboard deflection	3	3 h on surface
Turning circle (surfaced)	One rudder angle for each speed; repeated for port as well as starboard deflection	2	
Spiral manoeuvre in horizontal plane (surfaced)	One	3	
Grand Total			20 h

3.12.5 Analysis of the Trial Results

For the definitive manoeuvres given in Table 3.10 the analysis is fairly straightforward and the principal features can be obtained as required. Plots such as those shown in Figs. 3.59, 3.60, 3.61 and 3.62 can be generated.

However, for other trials where the objective is to generate data to correlate with free running model tests, or to validate numerical predictions, the task is more complex.

Firstly, it will be necessary to conduct any manoeuvring simulations with, as close as possible, the actual initial conditions as obtained during the full scale trials. In addition, any discrepancies in the ordered control surface angles, or the rpm which occurred in the full scale trials can be replicated in the simulation.

However, comparing real time data is fraught with difficulties, as if the measured and simulated paths deviate then it is difficult to determine why this is occurring, and to assess the relative accuracy of the simulations. For example, if after one minute the predicted depth is within the experimental error, but the pitch angle, and distance travelled is not, does this mean that the predictions are acceptable, or not?

In addition, if the control surface deflections were governed by an autopilot, then if the path and/or pitch/heading angle is different between the full scale trials and the simulation then the path deviation may be exaggerated (or otherwise)—again making it very difficult to assess the "accuracy" of the predictions from the simulation.

An alternative approach is to fix the simulation to follow the same path, speed, and pitch/heading angle as measured in the full scale trials, with the control surface deflections occurring at the same locations. Then, the additional external forces and moments required in the simulations to achieve this can be obtained. This will give a better measure of the accuracy of the simulation procedure. This also makes it easier to determine which aspects of the simulations need to be modified to result in an improved prediction.

References

Abbott IH, Von Doenhoff AE (1960) Theory of wing sections. Dover Publications, Inc. 1960. ISBN 10: 0486605868

Bayliss JA, Kimber NI, Marchant P (2005) Submarine trials and experimentation—dealing with real life data. In: Proceedings of warship 2005: naval submarines 2005, Royal Institution of Naval Architects

Bettle MC (2014) Validating Design Methods for Sizing Submarine Tailfins. In: Proceedings of warship 2014: naval submarines and UUVs, Royal Institution of Naval Architects

Bonci M (2014) Application of system identification methods for the evaluation of manoeuvrability hydrodynamic coefficients from numerical free running tests. Corso di Laurea Magistrale in Ingegneria Navale, Universita Degli Studi Di Genova, pp 19–21

Booth TB, Bishop RED (1973) The planar motion mechanism. Admiralty Experiment Works Publication

Broglia R, Di Mascio A, Muscari R (2007) Numerical study of confined water effects on self-propelled submarine in steady manoeuvres. Int J Offshore Polar Eng 17(2):89–96

Broglia R, Dubbioso G, Durante D, Di Mascio A (2015) Turning analysis of a fully appended twin screw vessel by CFD. part 1: single rudder configuration. Ocean Eng 105(2015):275–286

Crossland P (2013) Profiles of excess mass for a generic submarine operating under waves. In: Pacific 2013: international maritime conference, Sydney, Oct 2013

Crossland P, Marchant P, Thompson N (2011) Evaluating the manoeuvring performance of an X-plane submarine. In: Proceedings of warship 2011: Naval submarines and UUVs, Royal Institution of Naval Architects, Bath, 29–30 June 2011

Crossland P, Nokes RC, Dunningham S, Marchant P, Kimber N (2014) SRMII—A reconfigurable free running model capability for submarines with large L/D ratios. In: Proceedings of warship 2014: naval submarines and UUVs, Royal Institution of Naval Architects, Bath, 18–19 June 2014

Dempsey EM (1997) Static stability characteristics of a systematic series of stern control surfaces on a body of revolution, DTNSRDC Report 77-0085, Aug 1997

Dong PG (1978) Effective mass and damping of submerged structures. University of California, Lawrence Livermore Laboratory, Report No. UCRL-52342, California, Apr 1978

Dubbioso G, Muscari R, Di Mascio A (2013) Analysis of the performances of a marine propeller operating in oblique flow. Comput Fluids 75(2013):86–102

Dubbioso G, Muscari R, Ortolani F, Di Mascio A (2017) Analysis of propeller bearing loads by CFD. Part 1 straight ahead and steady turning manoeuvres. Ocean Eng 130(2017):241–259

Feldman J (1979) DTNSRDC revised standard submarine equations of motion. David W Taylor Naval Ship Research and Development Center, Ship Performance Department, DTNSRDC/SPD-0393-09, June 1979

Fox DM (2001) Small subs provide big payoffs for submarine stealth. Undersea Warfare 3(3)

Gertler M, Hagen GR (1967) Standard equations of motion for submarine simulation. Naval Ship Research and Development Center, Report No 2510, Washington, June 1967

Griffin MJ (2002) Numerical predictions of manoeuvring characteristics of submarines operating near the free surface. Ph.D. Thesis in Ocean Engineering at the Massachusetts Institute of Technology

Groves NC, Huang TT, Chang MS (1989) Geometric characteristics of DARPA Suboff models. David Taylor Research Center, SHD 1298-01, Maryland, USA, Mar 1989

Gutsche F (1975) The study of ships' propellers in oblique flow. Shiffbauforschung 3 3/4 (1964) pp 97–122 (from German), Defence Research Information Centre Translation No 4306, Oct 1975

Harris RG (1918) Forces on a propeller due to sideslip. ARC R & M 427

Haynes D, Bayliss J, Hardon P (2002) Use of the submarine research model to explore the manoeuvring envelope. In: Proceedings of warship 2002, Royal Institution of Naval Architects, June 2002

Itard X (1999) Recovery procedure in case of flooding. In: Proceedings of warship'99: naval submarines, Royal Institution of Naval Architects, June 1999, London

Jensen PS, Chislett MS, Romeling JU (1993) Den-Mark 1, an innovative and flexible mathematical model for simulation of ship manoeuvring. In: Proceedings of MARSIM'93, international conference on marine simulation and ship manoeuvrability, St John's

Jones DA, Clarke DB, Brayshaw IB, Barillon JL, Anderson B (2002) The calculation of hydrodynamic coefficients for underwater vehicles, Report Number: DSTO-TR-1329. DSTO Platforms Sciences Laboratory, Fisherman's Bend, Victoria, Australia

Korotkin AI (2009) Added mass of ship structure. Fluid mechanics and its applications, vol 88, Springer. ISBN 978-1-4020-9431-6

Landrini M, Casciola CM, Coppola C (1993) A nonlinear hydrodynamic model for ship manoeuvrability. In: Proceedings of MARSIM'93, international conference on marine simulation and ship manoeuvrability, St John's

Lloyd ARJM (1983) Progress towards a rational method of predicting submarine manoeuvers. In: Royal Institution of Naval Architects symposium on naval submarines, London

Lloyd ARJM, Campbell IMC (1986) Experiments to investigate the vortex shed from a body of revolution. In: 59th meeting of the AGARD fluid dynamics panel symposium, Monterey, Oct 1986

Lyons DJ, Bisgood PL (1950) An analysis of the lift slope of aerofoils of small aspect ratio, including fins, with design charts for aerofoils and control surfaces', ARC R&M No 2308

Mackay M (2001) Some effects of tailplane efficiency on submarine stability and maneouvring. Defence R&D Canada—Atlantic, Technical Memorandum, 2001-031, Canada, Aug 2001

Mackay M (2003) Wind tunnel experiments with a submarine afterbody model. Defence R&D Canada—Atlantic, Technical Memorandum, 2002-194, Canada, Mar 2003

Mackay M, Williams CD, Derradji-Aouat A (2007) Recent model submarine experiments with the MDTF. In: Proceedings of the 8th Canadian marine hydromechanics and structures conference, St John's, 16–17 Oct 2007

Marchant P, Kimber N (2014) Assuring the safe operation of submarines with operator guidance, UDT 2014: Liverpool, UK, 10–12 June 2014

Mendenhall MR, Perkins SC (1985) Prediction of the unsteady hydrodynamic characteristics of submersible vehicle. In: Proceedings of the 4th international conference on numerical ship hydrodynamics, Washington, pp 408–428

Molland AF, Turnock SR (2007) Marine rudders and control surfaces, principles, data, design and applications. Butterworth-Heinemann. ISBN: 978-0-75-066944-3

Musker AJ (1984) Prediction of wave forces and moments on a near-surface submarine. In: Shipbuilding: marine technology monthly, vol 31
Ortolani F, Mauro S, Dubbioso G (2015) Investigations of the radial bearing force developed during actual ship operations. Part 1: straight ahead sailing and turning manoeuvres. Ocean Eng 94(2015):67–87
Overpelt B (2014) Innovation in the hydrodynamic support for design of submarines. In: Proceedings of the 12th international naval engineering conference and exhibition 2014, Institution of marine Engineers, Scientists and Technologists, Amsterdam, 20–22 May 2014
Pitts WC, Nielsen JN, Kaarrari GE (1957) Lift and center of pressure of wing-body-tail combinations at subsonic, transonic, and supersonic speeds. NACA Report 1307:1957
Polis CD, Ranmuthugala D, Duffy J, Renilson MR, Anderson B (2013) Prediction of the safe operating envelope of a submarine when close to the free surface. In: Proceedings of the Pacific 2013 international maritime conference, Sydney, Oct 2013
Praveen PC, Krishnankutty P (2013) Study on the effect of body length on the hydrodynamic performance of an axi-symmetric underwater vehicle. Indian J Geo-Mar Sci 42(8):1013–1022
Pook DA, Seil G, Nguyen M, Ranmuthugala D, Renilson MR (2017) The effect of aft control surface deflection at angles of drift and angles of attack. In: Proceedings of warship 2017 naval submarines and UUVs, Royal Institution of Naval Architects, Bath, UK
Ray AV (2007) Manoeuvring trials of underwater vehicles: approaches and applications. J Ship Technol 3(2)
Ray AV, Singh SN, Sen D (2008) Manoeuvring studies of underwater vehicles—a review. In: Transactions of royal institution of naval architects, vol 150
Renilson MR, Ranmuthugala D, Dawson E, Anderson B. van Steel S, Wilson-Haffenden S (2011) Hydrodynamic design implications for a submarine operating near the surface. In: Proceedings of warship 2011: naval submarines and UUVs, Royal Institution of Naval Architects, 29–30 June 2011
Renilson MR, Polis C, Ranmuthugala D, Duffy J (2014) Prediction of the hydroplane angles required due to high speed submarine operation near the surface. In: Proceedings of warship 2014: naval submarines and UUVs, Royal Institution of Naval Architects, Bath, 18–19 June 2014
Ribner HS (1943) Formulas for propellers in yaw, and charts of the side force derivative. Report E319, Langley Memorial Aeronautical Laboratory, National Advisory Committee for Aeronautics, USA
Roddy RF (1990) Investigation of the stability and control characteristics of several configurations of the DARPA SUBOFF model (DTRC Model 5470) from captive-model experiments. David Taylor Research Center, SHD 1298-08, Maryland, USA, Sept 1990
Seil G, Anderson B (2013) The influence of submarine fin design on heave force and pitching moment in steady drift. In: Pacific 2013: international maritime conference, Sydney, Oct 2013
Sen D (2000) A study on sensitivity of manoeuverability performance on the hydrodynamic coefficients for submerged bodies. J Ship Res 44(3):186–196
Spencer JB (1968) Stability and control of submarines—Parts I–IV. J Roy Navy Sci Serv 23(3)
Sun S, Li L, Wang C, Zhang H (2018) Numerical prediction analysis of propeller exciting force for hull-propeller-rudder system in oblique flow. Int J Naval Archit Ocean Eng 10(2018):69–84
Tickle L, Bamford J, Hooper D, Philip N, Pinder G (2014) Sea trials and data analysis of the astute class submarine. In: Proceedings of warship 2014: naval submarines and UUVs, Royal Institution of Naval Architects, Bath, 18–19 June 2014
Tinker SJ (1988) A discrete vortex model of separated flows over manoeuvring submersibles. In: International conference on technology common to aero and marine engineering, society for underwater technology, Jan 1988
UCL (undated) Calculation of submarine derivatives. UCL course notes
Veillon A, Aillard JM, Brunet P (1996) Submarine depth control under waves: and experimental approach. In: Royal Institution of Naval Architects Warship'96: naval submarines 5—the total weapon system, London

Ward B (1992) Experiments to improve predictions of submarine manoeuvres. In: Proceedings of MCMC'92, Southampton, pp 248–260

Watt GD, Bohlmann H-J (2004) Submarine rising stability: quasi-steady theory and unsteady effects. In: Proceedings of the 25th symposium on naval hydrodynamics, St John's, Aug 2004

Whicker LF, Fehiner LF (1958) Free stream characteristics of a family of low-aspect-ratio all-movable control surfaces for application to ship design, David Taylor Model Basin Report Number 933, Dec 1958

Chapter 4
Resistance and Flow

Abstract The resistance of a submarine will have a major influence on its top speed, endurance, and acoustic signature. The various components of resistance include: surface friction; form drag; induced drag; and wave making resistance. The latter only becomes important when the submarine is operating on, or close to, the water surface. The flow over a submarine will influence its top speed, its acoustic signature, and the effectiveness of its own sensors. In particular, flow separation should be avoided. A submarine hull is usually considered in three parts: fore body; parallel middle body; and aft body. The main driver for the hydrodynamic design of the fore body is to control the flow such that there is laminar flow over the sonar array. A fuller fore body may be beneficial for this. The length of the parallel middle body influences the length to diameter ratio, and it is shown that there is an optimum value of the *L/D* to minimise resistance, depending on the hull form. The aft body shape can be characterised by the half tail cone angle, which defines its fullness. The primary aim of the design of the aft body is to avoid flow separation, and ensure good flow into the propulsor. Appendages contribute significantly to the hull resistance. In addition, they generate vortices which can have a detrimental effect on the flow around the hull, and in particular into the propulsor. Model testing and Computational Fluid Dynamics techniques are discussed. In addition, an empirical method of predicting the resistance of a submarine, suitable for use in the early stage of the design, is presented.

4.1 Introduction

The resistance (or drag) of a submarine will have a major influence on its top speed, and endurance. In addition, high resistance will affect the acoustic signature due to:

(a) increased flow noise; and
(b) the requirement for increased propulsion power to achieve a given speed.

M. Renilson, *Submarine Hydrodynamics*,
https://doi.org/10.1007/978-3-319-79057-2_4

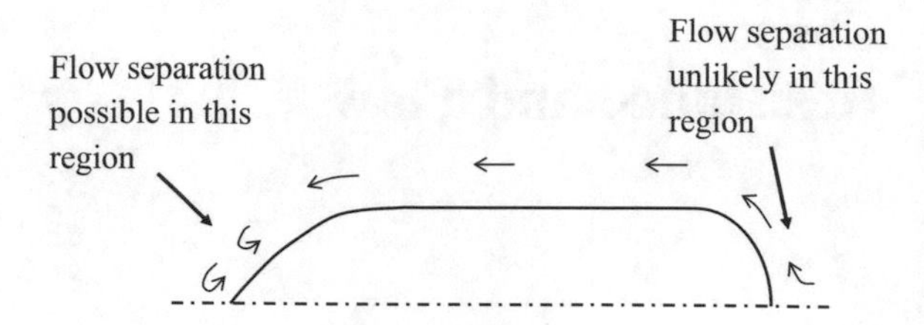

Fig. 4.1 Schematic of flow around a submarine

Water flow around the submarine will affect its self-noise, and hence the effectiveness of any sensors. Flow into the propulsor will affect both the propulsor noise, and the propulsor efficiency.

Flow separation occurs when the flow becomes detached from the surface of the hull, or appendage, and instead takes the form of eddies and vortices. This should be avoided.

Flow separation may occur when the cross section area of the submarine is decreasing along the length of the hull, and the flow is decelerating, as shown in Fig. 4.1. Thus, care needs to be taken whenever the cross sections are decreasing. This includes the region towards the stern, as shown in Fig. 4.1. However, it also includes any regions where casings, or other features, result in the cross sectional area reducing along the length of the hull, causing the flow to decelerate over the hull. On the other hand, flow separation is unlikely when the cross section area is increasing along the length of the hull and the flow is accelerating.

Thus, a comprehensive understanding of the water flow around the submarine is important to ensure that it is optimised, and that no serious issues are caused by an incorrectly designed flow regime.

4.2 Components of Resistance

In principle, the components of resistance for a submarine are similar to those for a surface ship, and are shown schematically in Fig. 4.2. Note that this figure is not to scale—the relative magnitude of the various components will depend on the hull shape, and on the proximity to the water surface.

The most important difference between the resistance of a surface vessel (or a submarine on the water surface) and a deeply submerged submarine is that the latter will have no wave resistance.

Thus, the submerged resistance will be equal to the sum of the following components:

(a) *Surface Friction as a Flat Plate*: This is equivalent to the friction of a flat plate with the same wetted area, and same length as the submarine, hence at the same Reynolds number.
(b) *Frictional-form Resistance*: Because the submarine has a shape to it, the flow velocity will not be the same as that over a flat plate. In some places it will be

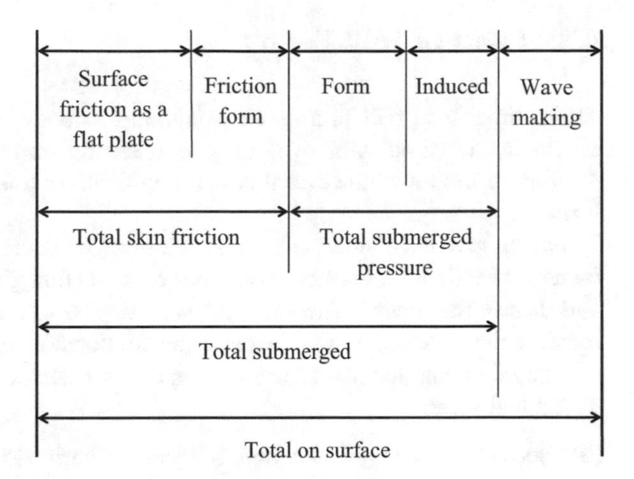

Fig. 4.2 Components of resistance

higher and in others it will be lower. The frictional-form resistance is the component of resistance caused by the difference in the flow between that on a flat plate and the actual flow over the submarine. This is generally a very small component of resistance.

(c) *Form Resistance, or Form Drag*: The form drag is the viscous pressure resistance due to the shape of the body. A "streamlined" form will have less form resistance than a blunt shape.

(d) *Induced Resistance, or Induced Drag*: The induced drag is the resistance caused by lift. This could be on appendages which are generating lift due to misalignment with the flow, or to the hull, which may be generating lift due to asymmetry.

The surface friction as a flat plate, and the frictional-form resistance together make up the skin friction resistance, and are the components of the resistance tangential to the hull.

The form drag and the induced drag the make up the total pressure resistance for a deeply submerged submarine, and are the components of the resistance normal to the hull. As the induced drag on a hull is usually very small, the total pressure resistance on the hull is very close to the form drag, and hence the term "form resistance" or "form drag" is often used to describe this component of resistance.

For a typical submarine the total submerged pressure resistance does not exceed 10–20% of the skin friction resistance. Thus, the skin friction dominates the resistance of a deeply submerged submarine. The best way to reduce this is to reduce the wetted surface area. Hence, streamlined forms which reduce pressure resistance at the expense of wetted surface, and hence frictional resistance, are generally not as attractive as at first appears.

In addition to the above components, when a submarine is operating on, or close to, the water surface it will generate surface waves. The wave making resistance is the resistance caused by the generation of these surface waves. This is a function of Froude number and the distance from the water surface.

4.3 Effect of Hull Form

The optimal bare hull form is an axisymmetric body with a longitudinal section similar to a teardrop, as shown in Fig. 4.3. The fullest section is approximately 30–40% aft of the bow, however this is not critical. This is known as a teardrop hull form.

For a given volume, increasing elongation decreases form resistance but increases frictional resistance. A circular cross section gives the lowest wetted area, and hence the lowest frictional resistance. Casings and other features which increase the wetted area will increase the frictional resistance.

The following two parameters are important when considering the overall shape of the hull form:

(a) slenderness ratio = *L/D*, where *L* is the submarine length, and *D* is its diameter; and
(b) prismatic coefficient, $C_p = \nabla / A_m L$, where ∇ is the volume of the submarine and A_m is its midships cross sectional area.

The drag coefficient is plotted as a function of *L/D* for an axisymmetric teardrop hull in Fig. 4.4, developed using CFD. The relative size of the skin friction resistance and the form drag can be seen from this figure.

This figure shows that for larger values of *L/D*, although the form drag is smaller, the skin friction resistance is greater.

Thus, the minimum resistance for an unappended teardrop hull occurs at a value of *L/D* approximately equal to 6.6. The optimum value of C_p is approximately 0.61. There is not a sharp trough to either of these curves, so having values slightly different from the optimal will not create a great penalty (Gertler 1950).

If the hull is not a teardrop form, but a hull form which incorporates a Parallel Middle Body (PMB) with a constant circular cross section for much of its length, then the relative values of the skin friction and form drag will change, as illustrated in Fig. 4.5. For this PMB hull form the optimum value of *L/D* is higher, approximately 8 (Leong et al. 2015; Crété et al. 2017).

The total drag on a PMB hull form is greater than that on a teardrop hull form. However, the construction costs of a teardrop hull form will be far higher than that of a conventional hull form. One alternative, often suggested, is to construct the pressure hull using a PMB circular cross section, and to clad this with a light cladding of teardrop shape, as shown in Fig. 4.6.

The shape shown in Fig. 4.6 has a lower drag coefficient than a PMB hull form, when based on midships section area, or volume. However, the mass of water in the

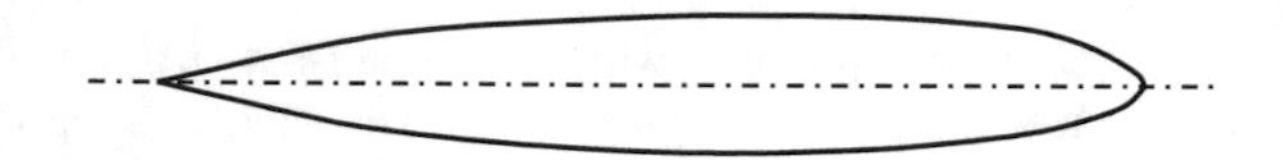

Fig. 4.3 Schematic of axisymmetric teardrop hull form

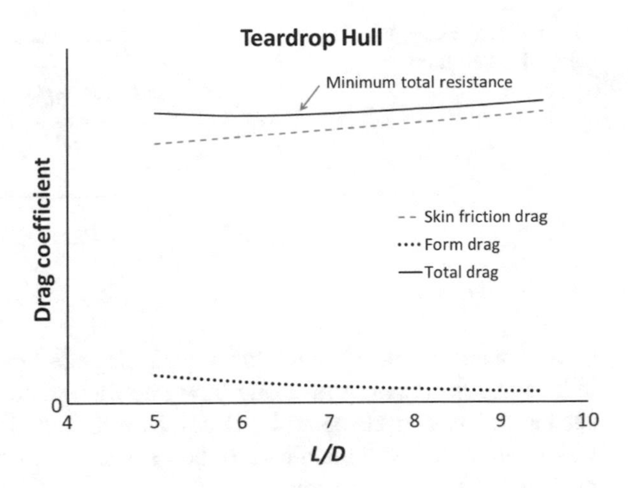

Fig. 4.4 Resistance components for an axisymmetric teardrop hull form

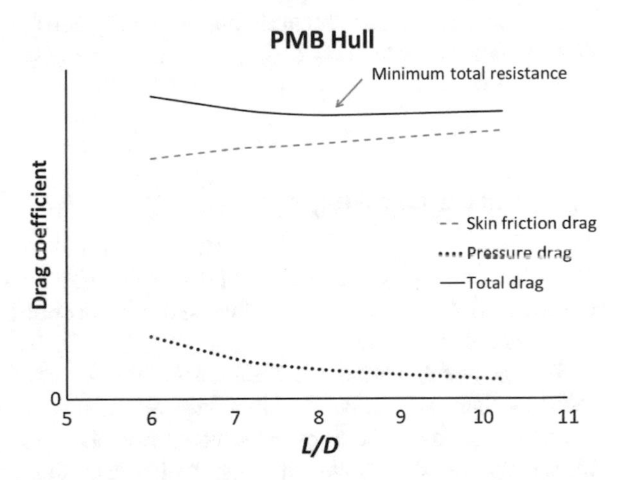

Fig. 4.5 Resistance components for a PMB hull form

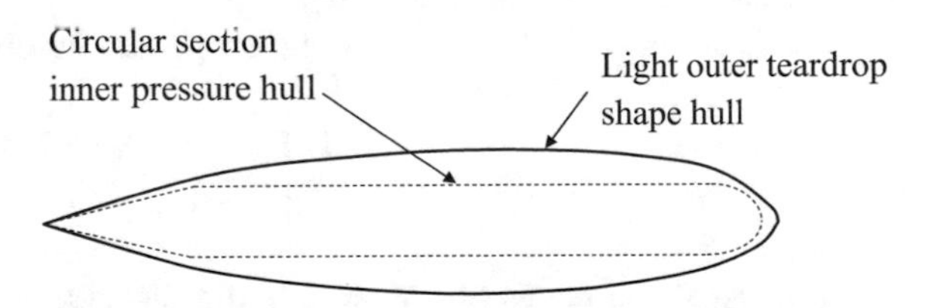

Fig. 4.6 Inner pressure hull with outer teardrop hull

free flood space between the pressure hull and the outer hull needs to be considered as part of the hull, and this additional mass is likely to outweigh the hydrodynamic advantage of the better hull form. This means that in many cases for a real submarine the simpler PMB hull form will actually give lower drag than the "optimized" teardrop hull form.

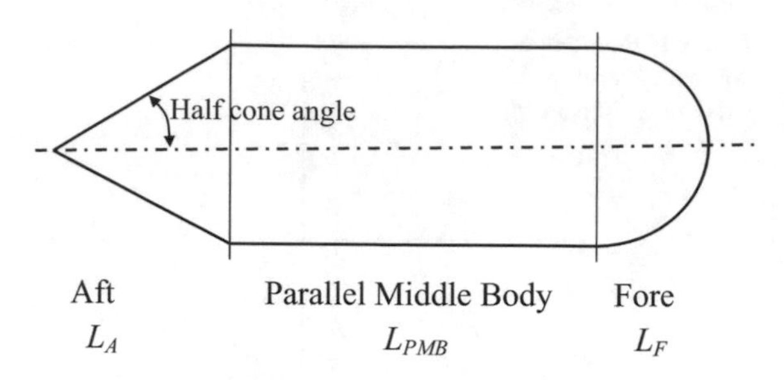

Fig. 4.7 Schematic of submarine hull form

An important consideration when analysing the flow around a submarine is to reduce any disturbed or unsteady flow into the propulsor. Thus, any appendages, such as sail, casing and control surfaces, have to be designed very carefully, and the wakes associated with them need to be tracked to ensure that they don't cause any problems with the propulsor.

For convenience, a submarine hull is usually considered in three parts: fore body (L_F); parallel middle body (L_{PMB}); and aft body (L_A). These are shown schematically in Fig. 4.7.

4.4 Fore Body Shape

The ideal fore body shape to minimise submerged resistance is an axisymmetric ellipsoid. However, changes to this optimum do not have a major impact on the submarine's resistance.

For many submarines the main design aim for the fore body shape is to control the flow such that the laminar flow regime extends as far aft as possible, to ensure laminar flow over the forward sonar array. This can generally be achieved by increasing the fore body fullness, which can delay transition by maintaining accelerating flow over the fore body.

If the fore body is axisymmetric its shape can be obtained from Eq. 4.1.

$$r_{x_f} = \frac{D}{2}\left[1 - \left(\frac{x_f}{L_F}\right)^{n_f}\right]^{\frac{1}{n_f}} \tag{4.1}$$

Here, r_{x_f} is the radius of the section at a distance x_f in the x-direction from the rearmost part of the fore body, as shown in Fig. 4.8. L_F is the length of the fore body, D is the hull diameter, and n_f is a coefficient which defines the fullness of the fore body. When $n_f = 1$ the bow profile is a conical form, and when $n_f = 2$ the bow profile is an elliptical form.

In general, the total resistance of the fore body will be greater for greater values of n_f, however the volume will also be greater, so the overall length of the submarine required to achieve the same buoyancy will be less. Thus, the value of n_f for

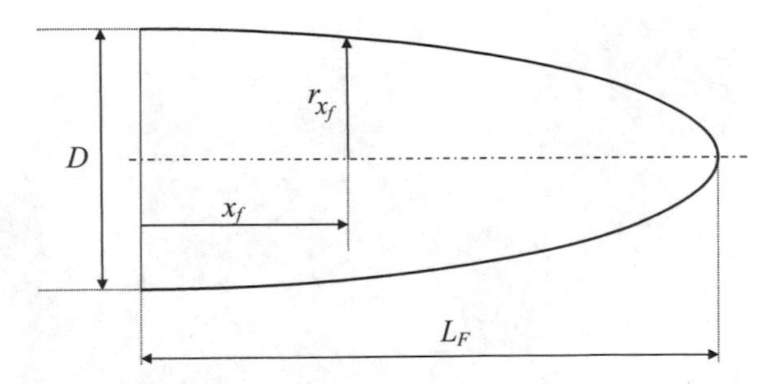

Fig. 4.8 Schematic of fore body

minimum total resistance should be chosen carefully. However, as noted above, the fore body shape is also used to control the flow such that the transition from laminar to turbulent occurs aft of the sonar array, if possible.

A value of n_f equal to about 2.2 gives a good fore body shape. In general, the pressure resistance of the fore body will be greater for greater values of n_f. Reducing n_f below 2.2 does not generally reduce pressure drag significantly. However, a larger value of n_f will result in a larger pressure drag on the fore body. Based on data from Moonesun and Korol (2017) the increased pressure drag on the fore body can be estimated by Eq. 4.2

$$\Delta C_{P_{fp}} = 0.01 \frac{A_F}{S} (n_f - 2.2) \tag{4.2}$$

where A_F is the fore body frontal area (=$\pi D^2/4$), and S is the wetted surface of the whole submarine. Note that Eq. 4.2 is valid for $2.2 \leq n_f \leq 5$.

An axisymmetric fore body as described above will generate significant wave making resistance when on the surface. For a submarine which spends limited time on the surface this is not considered to be a problem. However, for an SSK which may conduct significant transits on the surface in an overt (i.e. non-stealthy) mode, the increase in fuel consumption caused by the axisymmetric fore body can be significant. Thus, for such submarines a compromise may be required, with a fore body shape designed to take into account surface performance.

Figure 4.9a shows such a fore body shape, corresponding to a Type XXI U boat from the Second World War, whereas Fig. 4.9b shows the fore body shape on a Virginia Class submarine, a modern SSN. Work by Overpelt and Nienhuis (2014) has shown that a carefully designed fore body optimized for surface performance may reduce the resistance to 60% of that of a submarine with an axisymmetric fore body at high speeds, but have less than 5% increase in resistance when submerged. However, care will need to be taken regarding the influence of the fore body on the performance of the bow sonar.

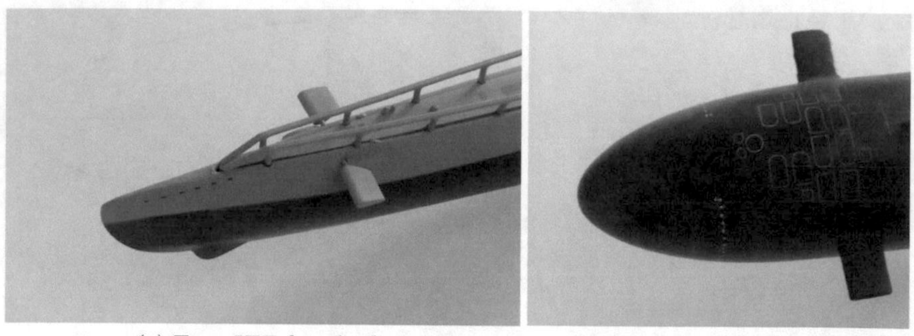

(a) Type XXI fore body (b) Virginia class fore body

Fig. 4.9 Models showing comparisons of different fore body shapes

4.5 Parallel Middle Body (PMB)

In order to minimise wetted surface area, the optimum cross section shape for the parallel middle body is circular. However, it is often desirable to deviate from this to provide a non-watertight lightweight casing on the top surface to make it easier for the crew to move about on the hull when the submarine is on the water surface, as shown in Fig. 4.10. Casings can also be used to store items which can then be accessed through hatches, however such items need to be able to withstand the Deep Diving Depth (DDD) of the submarine.

As shown in Fig. 4.5 the optimum *L/D* for a submarine with a PMB hull form is approximately 8. However, an increased *L/D* will result in a smaller diameter for a given volume, which can be cheaper to construct.

It is also worth bearing in mind that if the *L/D* value is too small then larger appendages may be required to maintain control. As appendages contribute substantially to the resistance (Chap. 6), then this may suggest that a larger value of *L/D* should be chosen.

Fig. 4.10 Model showing casing on modern submarine

4.6 Aft Body Shape

The principal feature of the aft body is the half tail cone angle, as defined in Fig. 4.7. Too long an aft body, with a low half cone angle, will result in increased wetted surface (hence greater skin friction resistance), increased weight and greater cost. On the other hand, if the aft body is too short, with a large half cone angle then it may cause the flow to separate which, in addition to increasing the self-noise and drag on the hull, will cause disturbed flow to enter the propulsor, and result in a considerable increase in propulsor noise.

The aft body needs to be considered in conjunction with the propulsor, as discussed in Sect. 5.1. For a single propulsor on the hull's axis the presence of the propulsor will result in an accelerating flow, as illustrated schematically in Fig. 4.11. As noted above, separation is less likely when the flow is accelerating. This means that the ideal form of the aft body without such a propulsor will be different to that with one. The presence of the propulsor will permit a higher half tail cone angle and a fuller aft body, as discussed by Warren and Thomas (2000).

The normal arrangement for a single propulsor on a modern submarine is with the control appendages ahead of the propulsor, as shown in Fig. 4.12. The axisymmetric arrangement also allows a large diameter low speed propulsor which means that the propulsor efficiency can be very high—see Chap. 5.

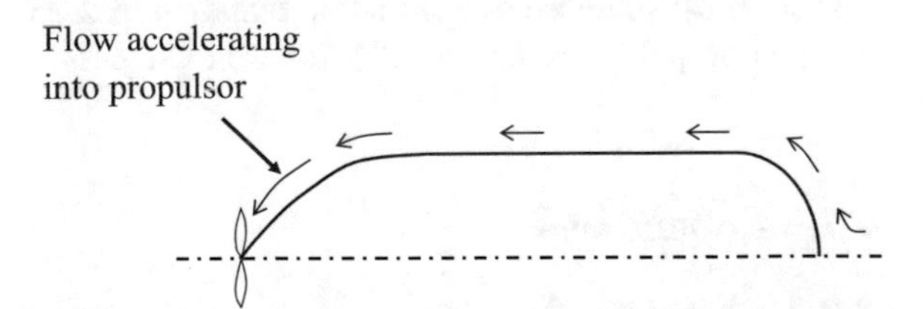

Fig. 4.11 Schematic illustrating flow being accelerated into propulsor

Fig. 4.12 Typical aft arrangement of propulsor and control appendages

In addition to the effect of the propulsor on the flow around the hull, the hull will affect the flow into the propulsor. This is due to:

(a) wake caused by the body ahead of the propulsor; and
(b) thrust deduction due to the low pressure ahead of the propulsor.

The hull efficiency, η_H, is the ratio of effective power to thrust power as discussed in sub Sect. 5.1.5. This is defined by Eq. 4.3.

$$\eta_H = \frac{(1-t)}{(1-w)} \tag{4.3}$$

where t is the thrust deduction fraction, and w is the Taylor wake fraction (Sect. 5.1). The full aft body will generate a high wake, which can result in hull efficiency values above 1.

The longitudinal position of the propulsor is not very critical to its performance, however it is desirable to have a reasonable separation between the propulsor and the appendages, as the disturbed flow from the appendages will create uneven flow into the propulsor, and generate propulsor noise.

In addition to accelerating the flow in the axial direction, a propulsor will also accelerate it in a rotational direction, which is lost energy, thus reducing its efficiency (sub Sect. 5.1.6). Surface ships often have the propeller immediately behind the rudder, thus reducing the lost energy due to rotational flow, but this is not the case for most submarines. Reducing rotational losses is one reason for adopting the pumpjet propulsor system, as discussed in Chap. 5.

4.7 Appendages

4.7.1 Introduction

Appendages contribute significantly to the drag of a submarine. In addition they cause vortices which generate noise, and can adversely affect the flow into the propulsor. This is particularly the case when the appendages are not aligned with the flow. Care needs to be taken with the alignment of eyebrow planes, as discussed in Sect. 6.3.3 and the aft control surfaces in an X-form configuration as discussed in Sect. 6.4.3.

Thus, careful consideration of the effect of appendages is important. Typical appendages include: the sail; bow and stern planes; rudders; and sonar domes. A sketch of a typical appendage in the hull boundary layer is shown in Fig. 4.13.

Notes on the design of the principal appendages are given in Chap. 6.

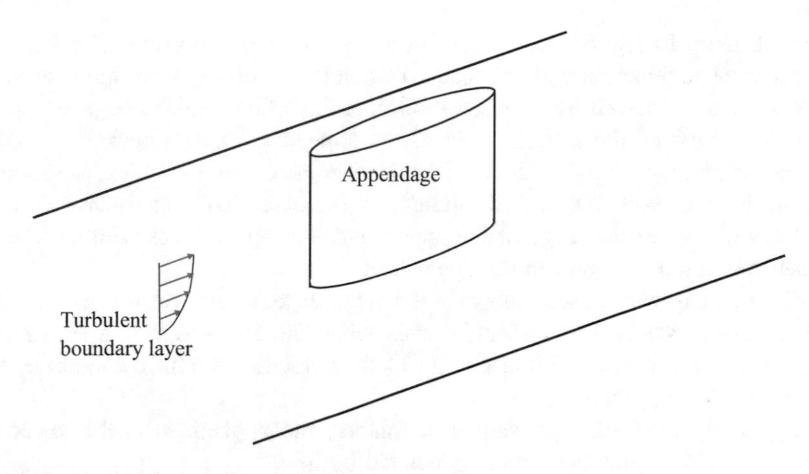

Fig. 4.13 Appendage in hull boundary layer

4.7.2 Appendages Aligned with the Flow

The interaction between a flow field along a surface and an obstruction (appendage) results in complex flow at the junction, as a result of the incoming flow field's boundary layer as shown in Fig. 4.14. This has implications in many areas and has been studied by a number of researchers, including: Stanbrook (1959), Hazarika and Raj (1987), Simpson (2001), Jones and Clark (2005), Olcmen and Simpson (2006), Fu et al. (2007), Coombs et al. (2012), and Toxopeus et al. (2014).

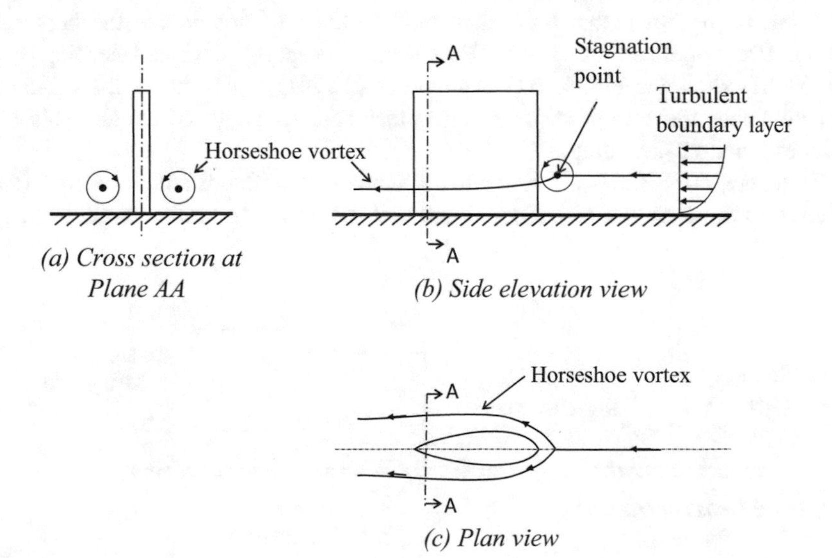

Fig. 4.14 Horseshoe vortex around an appendage

The primary feature of such a flow is the generation of a horseshoe vortex (also referred to as a "junction vortex"), caused when the incoming flow approaches the appendage, as discussed by Simpson (2001). A three dimensional stagnation point is formed ahead of the leading edge of the appendage, and a back flow occurs between this and the surface, ahead of the appendage, causing a vortex, as shown in Fig. 4.14b. The vortex size and strength is increased with an increase in the bluntness of the leading edge of the appendage. For appendages with sweepback the separation line is closer to the appendage.

The vortex is swept back alongside the appendage, as shown in Fig. 4.14.

The vortex structure is not stable, even when the appendage is a streamlined shape, and as discussed by Simpson (2001), the velocities can have a tendency to a bimodal velocity probability.

In general, the further forward, and thinner, the appendage is the lower the magnitude of the horseshoe vortex generated by it.

Because of the influence of the turbulent boundary layer on the behaviour of the vortex it is difficult to predict its behaviour using RANS, and hence it is often necessary to resort to more sophisticated computational techniques such as DES (Liu et al. 2010) or LES (Fureby et al. 2015).

In addition to the horseshoe vortex, secondary vortices can be formed, as shown in Fig. 4.15. Although only one of these secondary vortices is shown in Fig. 4.15, a number of these vortices can be generated under certain circumstances, as discussed in Simpson (2001).

The horseshoe vortices, and any associated secondary vortices, generate noise, and can influence the flow into the propulsor, further generating noise. (See for example Coombs et al. 2013). Thus, suppression of such vortices is important to reduce a submarine's noise signature.

It has been shown that a constant radius fillet will not reduce the horseshoe vortex (Devenport et al. 1990). However, a properly designed leading edge "strake", as shown in Fig. 4.16 (Devonport et al. 1991), that removes the separation can eliminate the vortex structure altogether at zero angle of attack. This also reduces junction flow drag.

Toxopeus et al. (2014) investigated a number of different strake designs (they referred to these as "cuffs"). They found that the optimum strake length to height

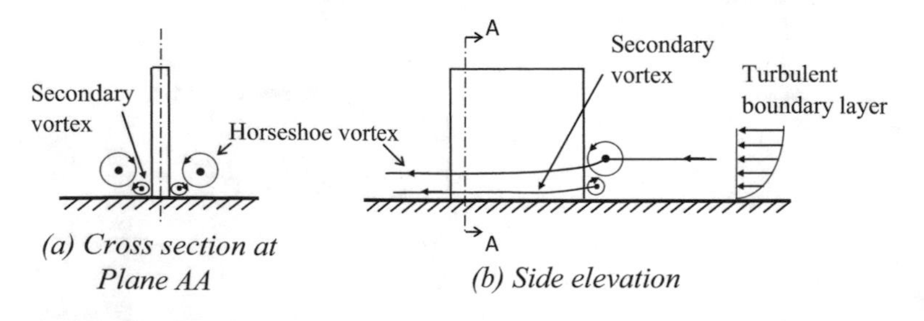

Fig. 4.15 Generation of secondary vortices

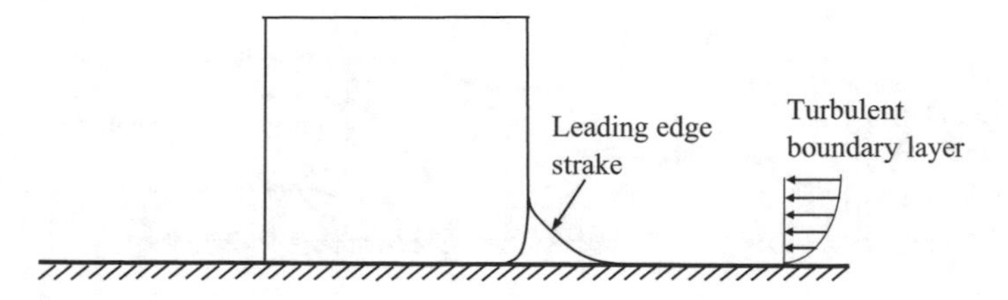

Fig. 4.16 Leading edge strake

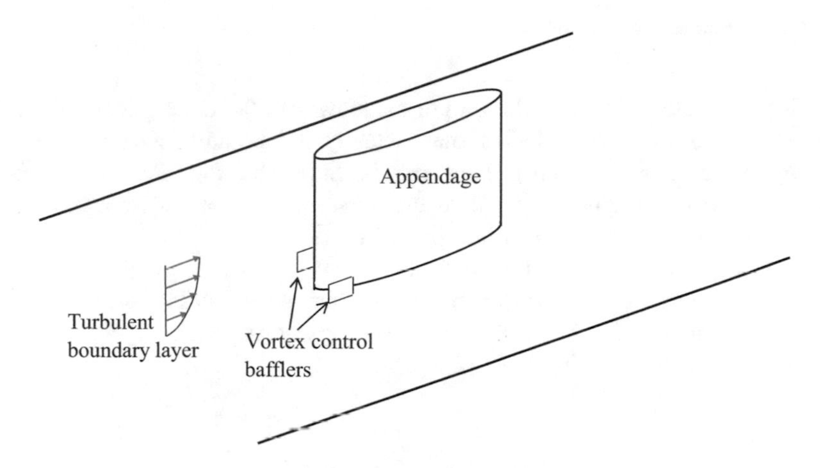

Fig. 4.17 Schematic of vortex control bafflers

was about two. Small strakes do not remove the horseshoe vortices, however strakes that are too large may increase the size of the tip vortices. The size of the strake also influences the quality of the flow into the propeller. For small strakes there is an improvement. However, if the strake is too large the wake is made worse. They found that the optimum size of the strake is related to the size of the sail, not the height of the boundary layer. They recommended a strake with a length of at least 18% of the length of the sail chord, and a height no greater than about 15% of the chord of the sail. This can add considerably to the volume of the sail.

Toxopeus et al. (2014) also concluded that the manoeuvring characteristics are not significantly influenced by the size of the strake.

Liu et al. (2010, 2011, and 2014) proposed the use of a "vortex control baffler" placed to the side of the leading edge of the appendix, as shown in Fig. 4.17. This is designed to generate an attached vortex with opposite rotation to the horseshoe vortex.

There is some evidence that such vortex control bafflers reduce the strength of the horseshoe vortex, and hence are beneficial to the inflow to the propulsor. Vortex control bafflers could be used in conjunction with the sail, and/or with the aft control surfaces, to improve the flow into the propulsor when the horseshoe vortex

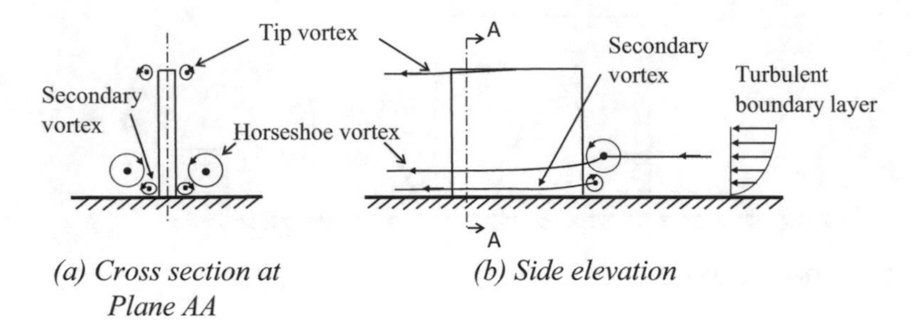

Fig. 4.18 Generation of tip vortices

that they generate impacts on the propulsor. However, the noise generated by the appendage plus the control baffler may actually not be any lower than the appendage on its own, and hence there may be little advantage in fitting a vortex control baffler to an appendage where the horseshoe vortex does not impact the propulsor—such as with a sail plane. See Chap. 6.

Tip vortices may also be formed from appendages at zero angle of attack, as discussed by Fureby et al. (2015), and shown schematically in Fig. 4.18. Toxopeus et al. (2014) found that if the tip of the sail is properly shaped the tip-vortex can be eliminated altogether at zero angle of attack.

4.7.3 Appendages at an Angle to the Flow

When appendages are operating at an angle to the flow the situation becomes more complex, as shown in Fig. 4.19 from Furbey et al. (2015). Note that in this figure the horseshoe vortex is referred to as "junction vortex".

For non-zero angles of attack the magnitude of the tip-vortex will be greater for greater angles, up until around 10°. Beyond that angle the magnitude of the tip-vortex may be less due to boundary layer separation (Jiménez and Smits, 2011).

Also, the strength of the horseshoe vortex is greater at greater angles of attack, up until an angle of around 15°–20°.

4.8 Operating Close to the Surface

4.8.1 Hull

When operating close to the water surface a submarine will generate surface waves, and hence wave resistance. This will depend on the Froude number (F_r) and the submergence of the submarine, (H^*).

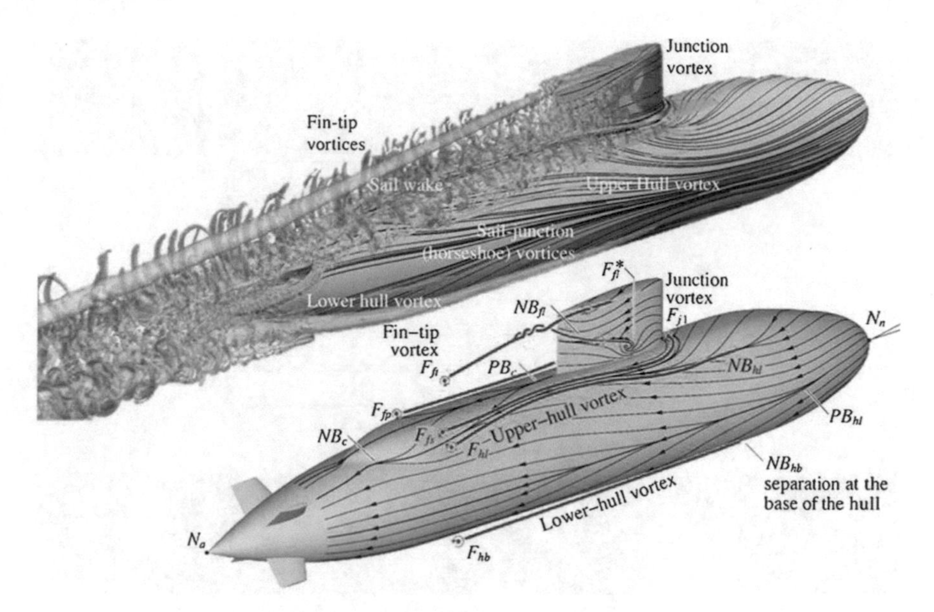

Fig. 4.19 CFD prediction of flow structure around submarine geometry at 10° yaw angle (reproduced with permission from DST Group)

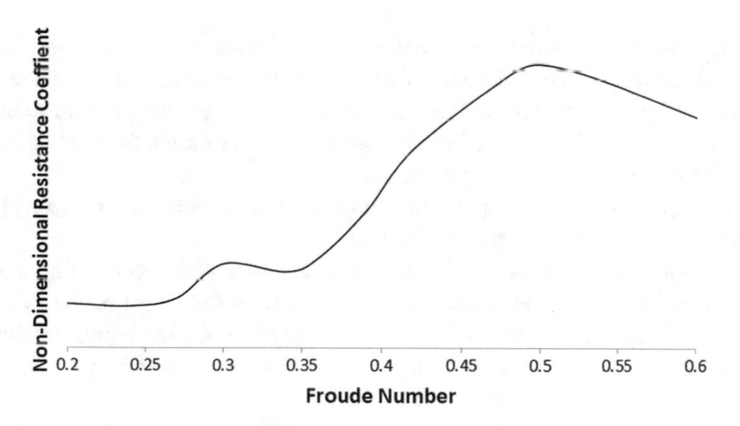

Fig. 4.20 Typical resistance curve showing effect of wavemaking resistance

A typical resistance curve showing non-dimensional resistance as a function of Froude number is given in Fig. 4.20. As can be seen, there are a number of humps and hollows due to the interference between the different wave patterns. A detailed discussion of wavemaking resistance is beyond the scope of this book, but can be found in naval architecture text books such as: Harvald (1983), Rawson and Tupper (2001), and Dern et al. (2016).

The important thing to note from Fig. 4.20 is that the locations of the peaks and troughs in the resistance curve are functions of Froude number. For a submarine

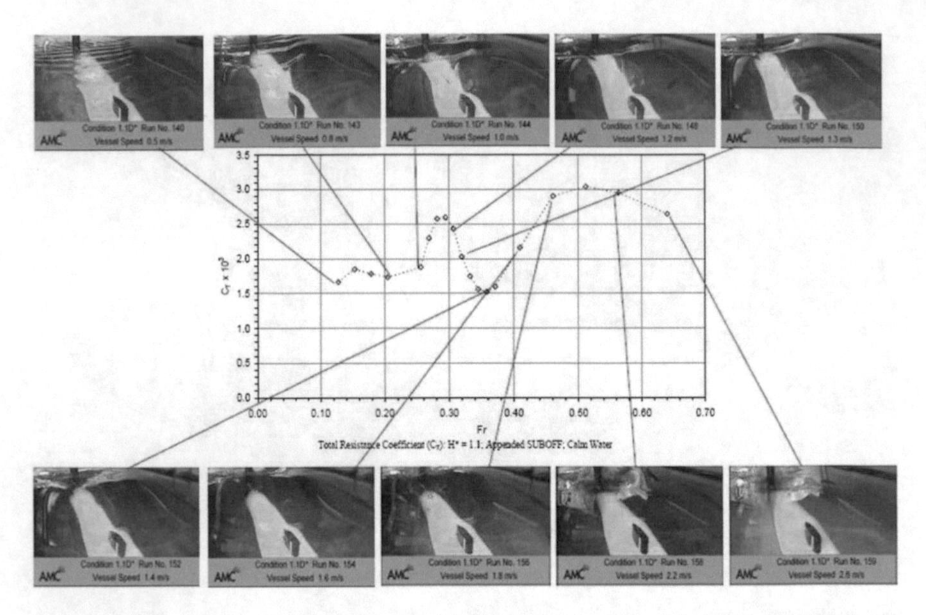

Fig. 4.21 Wave pattern and associated wave resistance for submarine close to the surface (courtesy of the Australian Maritime College)

travelling close to the surface these will be less pronounced than for a surface ship, however there will be the additional complication that individual wave patterns will be generated by the sail, and by the hull. As the lengths of these are different, for a given speed they will be operating at different Froude numbers. This adds to the complexity of the resistance curve.

An example of the wave pattern, along with the associated resistance, for a submerged submarine is given in Fig. 4.21.

As the sail will be close to the surface, and potentially generating the most significant wave pattern, the Froude number based on the chord of the sail should be considered, in addition to the Froude number based on the length of the hull.

Figure 4.22 shows a model running with its sail above the free surface, and the wave pattern generated can be clearly seen.

As wave resistance will not be able to be neglected, the optimum *L/D* ratio for a submarine operating close to the surface may be higher than one operating deeply submerged, (Renilson and Ranmuthugala 2012).

The volume in the sail may be a significant element, and should be considered for submarines which may spend considerable time close to the surface. Also, as noted in Sect. 4.4 Overpelt and Nienhuis (2014) have shown that the shape of the fore body for submarines designed to operate on the surface should be different to that for submarines which only operate submerged.

Further, the surface suction, as discussed in Sect. 3.8, will cause an out of plane force which will require deployment of appendages to compensate. This will result in induced drag on these appendages, which may need to be taken into account.

Fig. 4.22 Wave pattern generated by a submarine model with its sail above the surface (courtesy QinetiQ Limited, © Copyright QinetiQ Limited 2017)

When operating in wind generated waves there will be additional forces on the submarine which may need to be taken into account when determining its resistance.

4.8.2 Masts

When operating close to the surface masts such as periscope(s) and snorkels may be deployed. These will cause additional drag, as well as generate plume structures as shown in Fig. 4.23 from Conway et al. (2017).

The drag coefficient on a two dimensional circular cylinder without a free surface will depend on its Reynolds number. For Reynolds numbers in the range of interest this is likely to be between around 0.3 and 1.2 (Hoerner 1965). In addition, there will be the wavemaking resistance due to the wave plume as discussed in Conway et al. (2017b). For design purposes the use of a drag coefficient (based on frontal area) of 1.2 is likely to be appropriate for a mast with a circular cross section. Note that the drag coefficient can be significantly smaller for a streamlined mast.

In addition, if more than one mast is deployed there will be considerable interference between them. The share of the total drag will depend on which size of mast is ahead of the other, as can be seen in Fig. 4.24, where details of the four configurations are given in Table 4.1.

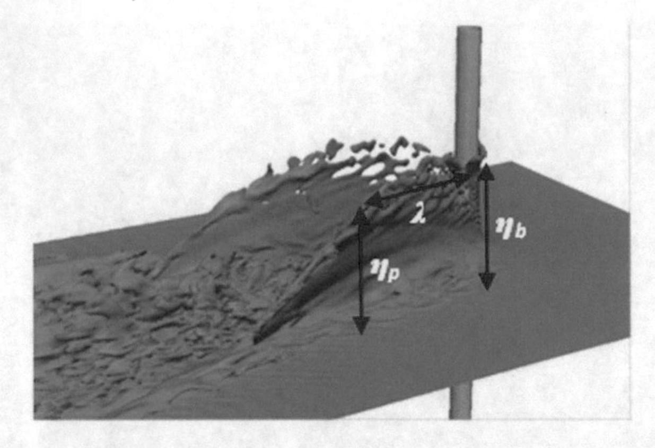

Fig. 4.23 Computer generated image of plume structure from single surface piercing mast (from Conway et al. 2017a)

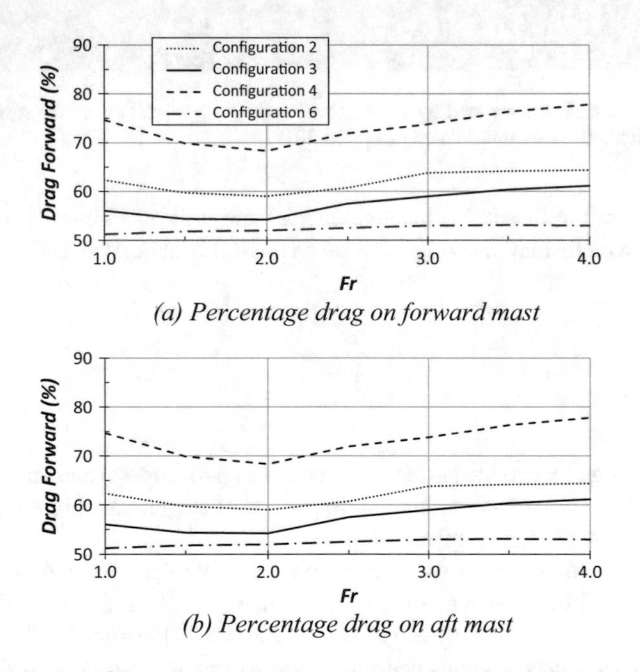

Fig. 4.24 Distribution of drag on two masts for various configurations in Fig. 4.1 (from Conway et al. 2017b). The Froude number is based on mast diameter

Table 4.1 Mast configurations for Fig. 4.24

Config. No.	Description
2	Two circular cylinders of equal area
3	Forward circular cylinder with smaller area
4	Forward circular cylinder with larger area
6	Two NACA0012 section cylinders with same area

In addition to the drag, surface piercing masts will also generate significant plume structures which will influence the visual signature, Conway (2017).

4.9 Prediction of Submarine Resistance

Predicting the resistance of a submarine is very important to assessing its operational capabilities, and an accurate knowledge of the resistance is necessary to design the propulsor and establish the power required. The resistance at non-ideal configurations, such as when not in hydrostatic balance, requiring the control surfaces to provide a vertical force, and when close to the surface snorkelling should also be considered.

The resistance of a submarine can be determined either by model testing, or by Computational Fluid Dynamics (CFD). For a deeply submerged submarine there will be no wave resistance, so the total resistance is made up of pressure resistance (primarily form drag) and frictional resistance. As noted in Figs. 4.4 and 4.5 the form drag is only a small component of the total. To estimate resistance it is necessary to consider full scale Reynolds number, if possible.

4.9.1 Model Testing

4.9.1.1 Turbulence Stimulation

As is the case when testing models of surface ships, the Reynolds number at model scale will be much lower than that at full scale so the flow over the model will be partly laminar. As with surface ship models, turbulent flow is achieved using turbulence stimulation close to the bow, and each hydrodynamic facility will have its own procedure for this. However, ensuring turbulent flow over the appendages, particularly the planes and rudders, is difficult since the Reynolds numbers over these can be quite small. Overly large turbulence stimulators will affect the flow and can result in excessive stimulator drag. This is one reason why large models are preferred, as even with 5 m long models the size of the planes and rudders are small enough to result in very low Reynolds numbers at speeds normally possible in a towing tank.

There are a number of different turbulence stimulation methods available including: trip wires; studs/pins; sand strips; and Hama Strips (Hama et al. 1957). A summary of different methods used for submarines is given in Erm et al. (2012).

The International Towing Tank Conference (ITTC) recommends procedures for turbulence stimulation suitable for surface ships, as given in ITTC (2011a). Note that this is updated from time to time by the ITTC, and is available on the ITTC website. Thus, reference should be made to the most recent version.

There is a good summary of turbulence stimulation procedures in the Resistance Committee report to the 28th ITTC (ITTC, 2017).

Work by Jones et al. (2013) on trip wires recommends that the Reynolds number based on the wire, Re_{d_T}, be in the range of 580–900 for effective transition. Values higher than 900 may over stimulate the flow. Re_{d_T} is defined by Eq. 4.4.

$$Re_{d_T} = \frac{U_1 d_T}{\upsilon} = \frac{U_\infty d_T \sqrt{1 - C_p}}{\upsilon} \tag{4.4}$$

where U_1 is the streamwise velocity at the edge of the boundary layer, d_T is the diameter of the trip wire, U_∞ is the nominal streamwise velocity, and C_p is the pressure coefficient.

Shen et al. (2015) developed a technique specifically for use with axisymmetrical bodies, such as submarines. They used a thin wire with wire size corresponding to $Re_{d_T} = 400$. The reasoning behind their approach was that the drag caused by conventional turbulence stimulators can be quite large. They showed that their turbulence stimulation technique generated a smaller drag, around 2% of the model resistance.

Different turbulence stimulation techniques are shown in Fig. 4.25.

Regardless of what turbulence stimulation method is used, it is important to remain consistent, such that the results from different tests can be compared. This is also important to make it possible to develop correlation allowances between model and full scale, as discussed in sub Sect. 4.9.1.4.

4.9.1.2 Resistance When Deeply Submerged

The normal method for predicting the resistance of a deeply submerged submarine is to test it in a towing tank as illustrated in Fig. 4.26. The model is supported by struts and tested inverted to reduce the interference caused by the struts. It is generally tested as deep as possible to avoid free surface effects, however it is also necessary to avoid any influence from the bottom of the tank. This approach allows for an operating propulsor if required for self-propulsion tests.

As it is not possible to achieve the full scale Reynolds number, and Froude number is not relevant to the resistance of a deeply submerged submarine, the procedure is generally to test at as high a speed as convenient, without generating significant surface waves. Most hydrodynamics facilities will have a standard speed at which they test.

Another approach for determining the resistance of a deeply submerged submarine is to make use of physical model tests in a wind tunnel. Depending on the size, and top speed available, it may be possible to obtain a higher Reynolds number in a wind tunnel than a towing tank. In addition, there are no complications with the generation of surface waves, and it is easier to make modifications to the model during the testing program. Flow visualisation is also easier to achieve. A photograph of a model being tested in a wind tunnel is given in Fig. 4.27.

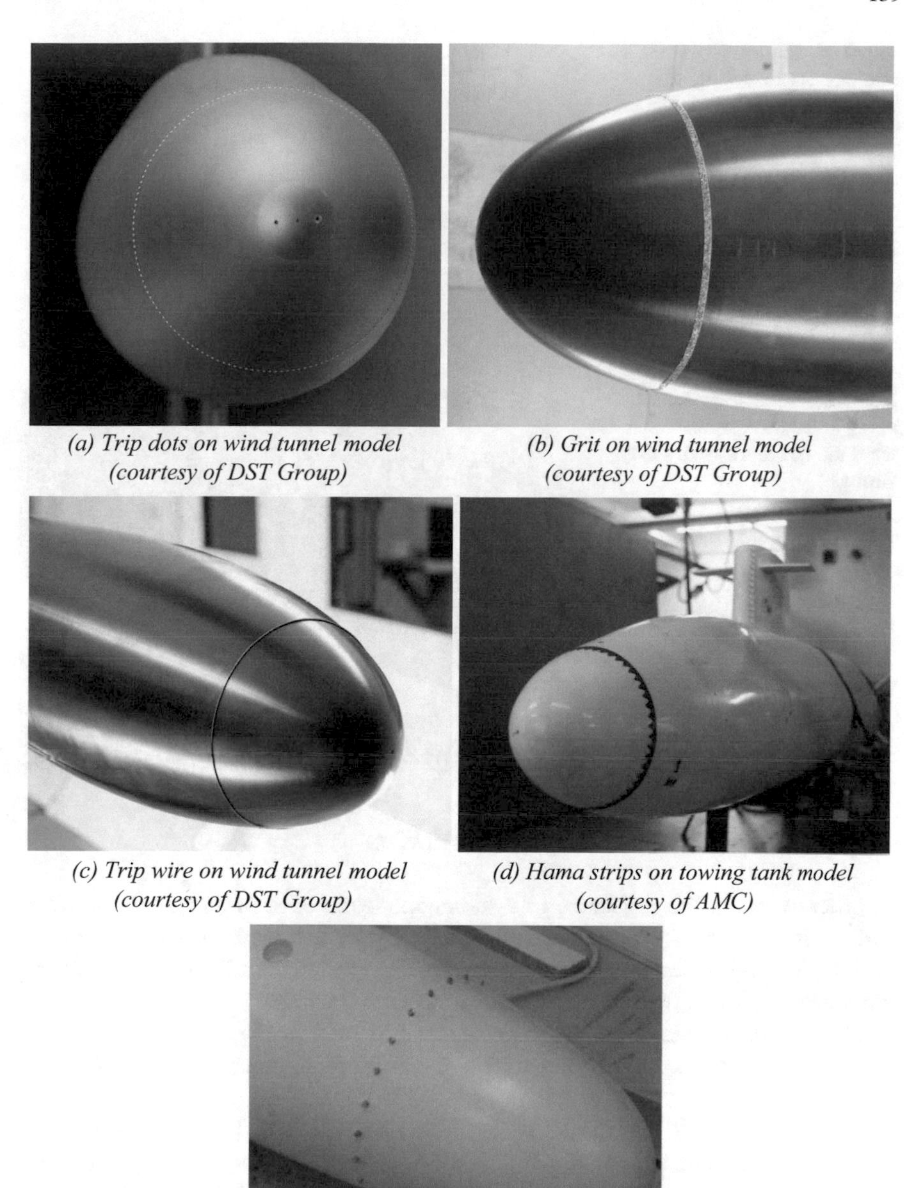

(a) Trip dots on wind tunnel model (courtesy of DST Group)

(b) Grit on wind tunnel model (courtesy of DST Group)

(c) Trip wire on wind tunnel model (courtesy of DST Group)

(d) Hama strips on towing tank model (courtesy of AMC)

(e) Trip studs on towing tank model (courtesy of AMC)

Fig. 4.25 Turbulence stimulation

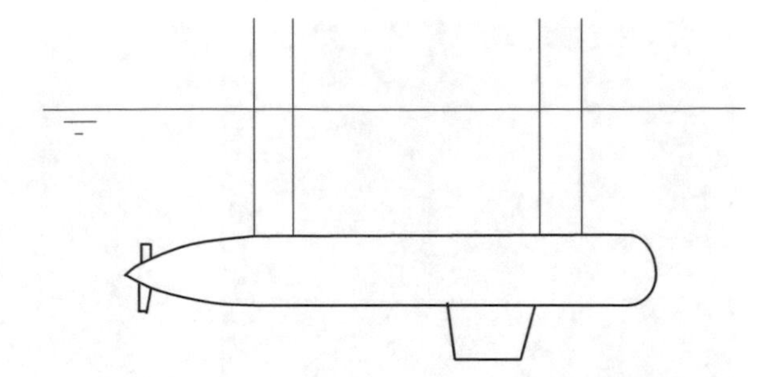

Fig. 4.26 Typical arrangement for testing the resistance of a submarine in a towing tank

Fig. 4.27 Submarine model being tested for resistance in a wind tunnel (courtesy of DST Group)

4.9.1.3 Resistance Close to the Water Surface

In order to obtain the resistance of a submarine when operating close to the water surface it is necessary to conduct model tests in a towing tank with the model upright. Testing inverted is no longer applicable as interaction between the sail and the water surface is one of the aspects being investigated.

This can cause additional problems. As can be seen in Fig. 4.28, using the two strut approach when close to the surface could result in considerable interference with the sail.

Alternatives are to use the aft sting support, as shown in Fig. 4.29, or a single strut through the sail, as shown in Fig. 4.30. Note that in both cases great care needs to be taken to ensure that the test rig is sufficiently stiff. This is particularly the case with the aft sting support. With the aft sting support it is not possible to test with an operating propulsor. As noted in Sect. 4.6 a propulsor can significantly affect the flow over the aft body, meaning that the optimum shape without a propulsor is different to that with a propulsor. However, this technique may be adequate for determining the influence of the proximity to the water surface on the resistance of the submarine, as this is probably not affected by the presence of the propulsor.

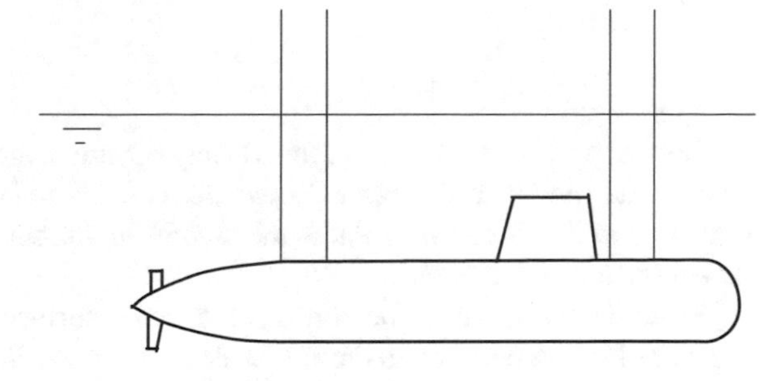

Fig. 4.28 Schematic of submarine being tested close to the surface using the two strut system

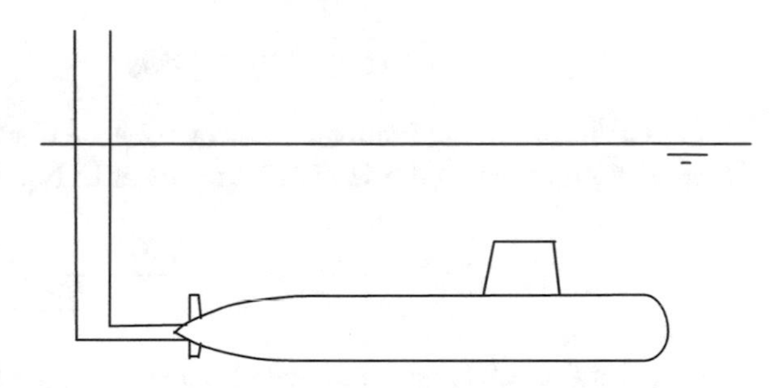

Fig. 4.29 Schematic of submarine being tested close to the surface using an aft sting support

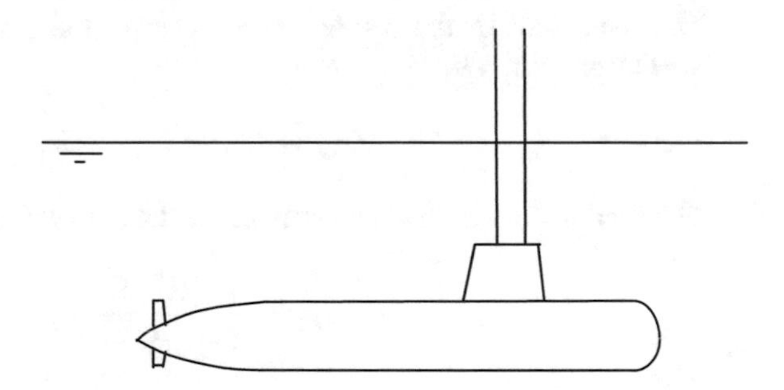

Fig. 4.30 Schematic of submarine being tested close to the surface using a single strut support system

4.9.1.4 Model to Full Scale Prediction

Scaling of resistance

The classical approach to scaling resistance results from model tests to full scale is to simplify the total resistance into the frictional component and the remainder, known as the residual resistance. It is assumed that the coefficient of frictional resistance is a function of Reynolds number, given by Eq. 4.5, and the coefficient of residual resistance is a function of Froude number, given by Eq. 4.6.

$$R_e = \frac{VL}{\upsilon} \tag{4.5}$$

$$F_r = \frac{V}{\sqrt{gL}} \tag{4.6}$$

From Eqs. 4.5 and 4.6 it is immediately apparent that it is not possible to test a small scale model at the same Reynolds number *and* Froude number as the full scale ship at the same time, unless the kinematic viscosity, υ, of the testing fluid can be substantially decreased.

As this is not possible, the normal practice for surface ships is to test the model at the same Froude number as the full scale ship so that the non-dimensional residual resistance component for the model is the same as that for the ship, as given in Eq. 4.7.

$$C_{R_M} = C_{R_S} \tag{4.7}$$

The coefficient of the frictional resistance for the model can be estimated using a correlation line such as the ITTC'57 line given in Eq. 4.8.

$$C_{F_M} = \frac{0.075}{(\log_{10} R_{e_M} - 2)^2} \tag{4.8}$$

Note that Eq. 4.8 is not a flat plate friction line, but makes a small allowance for the three dimensional effect of conventional surface ships.

The coefficient of the residual resistance for the model (and the ship) can be obtained from Eq. 4.9.

$$C_{R_M} = C_{R_S} = C_{T_M} - C_{F_M} \tag{4.9}$$

The frictional resistance component for the ship can be obtained from Eq. 4.10.

$$C_{F_S} = \frac{0.075}{(\log_{10} R_{e_S} - 2)^2} \tag{4.10}$$

Hence the total resistance coefficient for the ship can be obtained from Eq. 4.11.

$$C_{T_S} = C_{F_S} + C_{R_S} \tag{4.11}$$

This simplified approach does not specifically take into account the additional components of resistance, as shown in Fig. 4.2. Thus, the ITTC developed a slightly different approach, the ITTC'78 method (ITTC 2011b), which allows for a hull shape dependent form factor, *k*. With this method, instead of assuming that the total resistance is made up of the frictional component and the residual component, an additional form resistance is allowed for. This form resistance lumps both the friction form and the form drag illustrated in Fig. 4.2 into one component as given by Eq. 4.12.

$$C_{T_M} = (1 + k)C_{F_M} + C_{R_M} \tag{4.12}$$

For surface ships a number of different approaches have been developed for determining the form factor, *k*, as a function of hull shape. One such approach is to test at very low Froude numbers, where the residuary resistance is low, and extrapolate to zero Froude number. Another approach is to test a deeply submerged double body model, where the model hull below the waterline is mirrored, such that the flow over the hull is not influenced by the water free surface.

For submarines, the form factor can be obtained directly from testing a deeply submerged model, as the residuary resistance is zero for a deeply submerged submarine.

Roughness and correlation allowance

Model tests are generally conducted on smooth models in the controlled environment in a towing tank.

To account for the difference between the towing tank and real life an additional allowance has been developed by the ITTC (ITTC 2011b). This additional allowance has two components:

i. roughness allowance, ΔC_F; and
ii. correlation allowance, C_A.

An older approach is to adopt a single value of 0.0004 for the total of both components. However, a more sophisticated approach is to use different methodologies for each. This relies on making use of a large data base of ships where both the model and full scale values are available. This information is specific to ship types, and also to the particular towing tank facility.

Additional complications when testing submarine models are:

i. strut interference on the model tests;
ii. wave resistance when testing in a towing tank, even when deeply submerged;
iii. large appendages with laminar flow;
iv. blockage, which will be different to that experienced for surface ship models; and
v. extra roughness due to vent holes and other small protuberances.

Unfortunately the number of submarines tested at model scale, and subsequently trialled at full scale, is quite limited, and consequently it is not so easy to develop a relevant correlation allowance.

Thus, it is difficult to establish suitable values of ΔC_F and C_A for submarines.

4.9.2 Computational Fluid Dynamics (CFD)

State of the art CFD techniques can be used to estimate the resistance of a submarine, either deeply submerged, or close to the surface. As the resistance of a deeply submerged submarine is dominated by the frictional component, there are a number of difficulties with this, in particular the choice of empirically based turbulence model. However, in principle it is possible to use CFD to obtain results at full scale Reynolds numbers, something which is not possible using model experiments.

CFD can be used very effectively to study the flow regime, and in particular the flow patterns into the propulsor as a result of wake from appendages.

CFD can also be used effectively to determine the effect of small changes in the hull form. However, one of the current complications with CFD is that there is no standard method for predicting submarine resistance. This is largely because both computing power, and CFD techniques, are developing rapidly. Thus, great care needs to be taken when investigating the effect of the change in resistance due to a change in hull shape, compared to the original hull shape, possibly developed a few years previously. Improvements in CFD will likely mean that, unless care is taken, the method (grid size, turbulence model, y^+, etc.) used for the new hull form is likely to be different to that used for the original hull form. If this were to occur the difference in the result could be just as likely to be due to the new CFD technique, as to the new hull shape. This means that it is necessary to ensure that the resistance is obtained from both hull shapes using the same CFD procedure.

Finally, as with physical model testing, there is also uncertainty regarding any correlation allowance that needs to be applied from the results of the CFD to the full scale prediction. This is probably even more difficult to obtain with CFD than model tests, as there is no standard CFD procedure.

4.9.3 Approximation Techniques

Sometimes it is necessary to be able to make a prediction of the resistance of a submarine at the concept stage, prior to the commissioning of any model tests or CFD, both of which are rather expensive.

One approach is to consider the various elements individually, and then combine them. It may then also be necessary to add an additional allowance for interference between them.

Care needs to be taken when adding the individual elements as the non-dimensionalisation method is different for each.

4.9.3.1 Hull

As discussed in Sect. 4.2 the primary components of resistance on the hull are the skin friction; and the form drag. The flat plate friction can be obtained from the Hughes flat plate friction line given in Eq. 4.13.

$$C_{F_{flat}} = \frac{0.067}{(\log_{10}R_e - 2)^2} \tag{4.13}$$

Here $C_{F_{flat}}$ is the non-dimensional flat plate frictional resistance, defined by Eq. 4.14, and R_e is the Reynolds number based on the length of the submarine.

$$C_{F_{flat}} = \frac{R_{F_{flat}}}{\frac{1}{2}\rho S V^2} \tag{4.14}$$

$R_{F_{flat}}$ is the flat plate frictional resistance, S is the wetted surface of the hull, and V is the velocity. Note that when calculating the wetted surface the influence of the sail should be taken into account, and the sail footprint removed from the calculation of the wetted surface of the hull. For foil style sails (see Sect. 6.2) this may not be important, however for blended style sails it should be taken into account, as it can make a significant difference (Seil and Anderson 2012).

If the wetted surface of the submarine is not known (at an early design stage) this can be estimated using Eq. 4.15.

$$S_{hull} \approx 2.25D(L - L_{PMB}) + \pi D L_{PMB} \tag{4.15}$$

The first term in Eq. 4.15 is a good approximation for the wetted surface of the total of the fore body plus the aft body, whereas the second term is the addition for the parallel middle body. Note that this is for a circular cross section and that if a casing is fitted it will need to be increased accordingly.

The hull wetted area due to the footprint of the sail can be deducted from this.

However, as discussed in Sect. 4.2, there is an additional friction component due to the form of the vessel, which results in the flow velocity over the hull being different to that over a flat plate. It is very difficult to estimate what this is, however for surface ships the ITTC'57 line is often used, as given in Eq. 4.16.

$$C_{F_{form}} = \frac{0.075}{(\log_{10}R_e - 2)^2} \tag{4.16}$$

$C_{F_{form}}$ is the non-dimensional frictional resistance including the frictional-form resistance component for a surface ship.

However, recent CFD data for three axisymmetrical hull forms (Leong 2017) has shown that a small additional allowance to Eq. 4.16 is required to account for the effect of L/D on the friction-form resistance. Thus, for submarines a good estimation of the total non-dimensional frictional resistance can be given by Eq. 4.17.

$$C_{F_{form}} = \frac{0.075}{(\log_{10} R_e - 2)^2}[1 + K_F] \tag{4.17}$$

where

$$K_F = 0.3\frac{D}{L}$$

The factor, K_F accounts for the friction-form resistance of the axisymmetric hull form.

In addition to this there needs to be a further allowance due to the fact that the submarine hull is not smooth, but has numerous imperfections on it, ranging from general unevenness in the hull/casing, to vent holes for the ballast tanks. As very little data exists for correlation, it is difficult to estimate this component. For surface ships an additional value of 0.0004 is sometimes added to the non-dimensional frictional resistance to allow for hull roughness, however this does not take into account the additional drag due to the vent holes for the ballast tanks etc. See sub Sect. 4.9.1.4.

The form drag, or pressure drag, is a much smaller component of the drag on the submarine hull. The viscous pressure resistance can be obtained from Eq. 4.18.

$$C_P = K_P C_{F_{form}} \tag{4.18}$$

K_P is given by Eq. 4.19.

$$K_p = \left[\xi_{hull} + \xi_{PMB}\left(\frac{L_{PMB}}{L}\right)^{n_{PMB}}\right]\left(\frac{L}{D}\right)^n \tag{4.19}$$

where L_{PMB} is the length of the parallel middle body, and the values of the constants can be obtained from Table 4.2.

In Eq. 4.19 the first term in the square brackets represents the pressure resistance on a hull with no parallel middle body (such as those in the Series 58, Gertler 1950) and the second term accounts for parallel middle body. The values of the constants given in Table 4.2 have been obtained from data provided by Leong (2017).

Table 4.2 Values of constants for Eq. 4.19

ξ_{hull}	4
ξ_{PMB}	15
n_{PMB}	3
n	−1.8

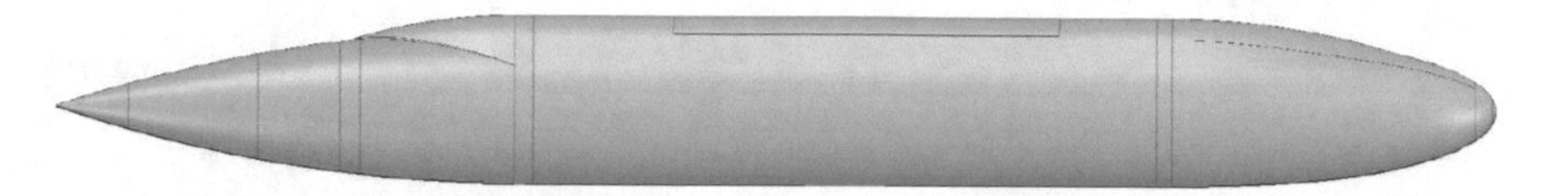

Fig. 4.31 Profile of submarine showing casing

Table 4.3 Casing factor

Type of casing	Casing factor, K_c
No casing	1.0
Simple casing, as Fig. 3.2	1.15

For a full fore body, where n_f is greater than about 2.2 (see Fig. 4.8), the pressure resistance of the fore body will be greater than estimated using Eq. 4.18. The increased pressure drag on the fore body can be estimated from Eq. 4.2.

This is for an axisymmetric body. An additional allowance may need to be made for a casing, or other modifications to the simple shape, if necessary.

The friction resistance for the casing can be obtained in the same way as for the hull. The additional casing wetted area should be used.

Care may need to be taken regarding additional vents, hatches, and other imperfections on the casing as these all increase the friction resistance.

The pressure resistance on the casing will depend on its shape. However, for a typical casing, as shown in Fig. 4.31, the increase in pressure resistance is of the order of 15% of that of an axisymmetrical shape.

The total resistance of the hull can be obtained from Eq. 4.20.

$$R_{T_{hull}} = \frac{1}{2}\rho V^2 S_{hull}\left[C_{F_{form_{hull}}} + \Delta C_{F_{form_{hull}}} + K_c C_{P_{hull}} + \Delta C_{P_{bow}}\right] \tag{4.20}$$

where S_{hull} is the wetted surface of the hull, including the casing, but excluding the footprint of the sail, and K_c is the casing factor as given in Table 4.3.

At this stage, values of the casing factor for other casing designs are not available. An estimate can be made as required. However, it is important to recognise that this will not have a major effect on the total resistance of the submarine.

4.9.3.2 Sail

As with the hull, the primary components of resistance on the sail are the skin friction, and the form drag.

The friction resistance coefficient can be obtained in much the same way as the friction resistance coefficient of the hull using Eq. 4.21.

$$C_{F_{flat}} = \frac{0.067}{(\log_{10} R_e - 2)^2} \tag{4.21}$$

where R_e is the Reynolds Number based on the chord of the sail.

Equation 4.21 is for a flat plate. CFD data from Leong (2017) shows that a good prediction is obtained by using Eq. 4.22. This applies for sail thicknesses from 15 to 40%.

$$C_{F_{form_{sail}}} = \frac{0.08}{(\log_{10} R_e - 2)^2} \tag{4.22}$$

Here, R_e is the Reynolds Number for the sail.

In addition to this, there needs to be a further allowance due to the fact that the sail is not smooth, but has numerous imperfections on it, referred to as $\Delta C_{F_{form_{sail}}}$.

As very little data exists for correlation, it is difficult to estimate this component. For surface ships, an additional value of 0.0004 is sometimes added to the non-dimensional frictional resistance based on surface area to allow for roughness. Even so, this still does not take into account the additional drag due to the vent holes etc.

The pressure drag on the sail can be estimated from Eq. 4.23 based on data from Leong (2017), again for a range of sail thicknesses from 15 to 40%.

$$C_{P_{sail}} = 10\left(\frac{t_{sail}}{c_{sail}}\right)^{1.75} C_{F_{form_{sail}}} \tag{4.23}$$

The total resistance of the sail can be obtained from Eq. 4.24.

$$R_{T_{sail}} = \frac{1}{2}\rho V^2 S_{sail}\left[C_{F_{form_{sail}}} + \Delta C_{F_{form_{sail}}} + C_{P_{sail}}\right] \tag{4.24}$$

where S_{sail} is the wetted surface of the sail.

4.9.3.3 Control Surfaces

For a thin streamlined shape with its maximum cross section approximately 30% aft of the leading edge, the total drag can be obtained using data from Hoerner (1965), modified to take into account data provided by Leong (2017). The combination is Eq. 4.25.

$$C'_{T_{cs}} = \left[2 + 8\left(\frac{t_{cs}}{c_{cs}}\right) + 120\left(\frac{t_{cs}}{c_{cs}}\right)^{4.5}\right] C_{F_{form_{cs}}} \tag{4.25}$$

where $C'_{T_{cs}}$ is the drag coefficient based on planform area, t_{cs} is the control surface thickness and c_{cs} is the control surface chord. $C_{F_{formcs}}$ is the friction, including friction form, coefficient given by Eq. 4.26.

$$C_{F_{formcs}} = \frac{0.08}{(\log_{10} R_e - 2)^2} \tag{4.26}$$

Here, R_e is the Reynolds Number for the control surface.

Equation 4.25 has been developed to use planform area of the control surface, rather than wetted surface area of the control surface, as the latter may not be known at an early stage of the design.

The resistance of the control surface is obtained from Eq. 4.27.

$$R_{T_{cs}} = \frac{1}{2}\rho V^2 A_{plan} C'_{T_{cs}} \tag{4.27}$$

where A_{plan} is the planform area of the control surface, and $C'_{T_{cs}}$ is obtained from Eq. 4.25.

The total resistance of all the control surfaces can be obtained by summing the values for each control surface.

4.9.3.4 Total Resistance

The total predicted resistance on the submarine can be obtained from Eq. 4.28.

$$R_T = R_{T_{hull}} + R_{T_{sail}} + \sum R_{T_{cs}} \tag{4.28}$$

4.9.3.5 Calculation of Resistance for Typical Submarine

The resistance of an example submarine geometry with principal parameters in Table 4.4 can be estimated using the above approximation technique.

It is assumed that there are four identical aft control surfaces and two identical forward control surfaces.

The water density is 1024.7 kg/m^3 and the kinematic viscosity is 1.05372×10^{-6} m^2/s.

The submarine speed is 5 m/s.

The values of the various components of resistance are given in Table 4.5, and presented graphically in Fig. 4.32.

Table 4.4 Principal particulars of example submarine geometry

Parameter	Symbol	Value
Length	L	70 m
Length of fore body	L_F	7 m
Length of parallel middle body	L_{PMB}	50 m
Length of aft body	L_A	13 m
Diameter	D	6.5 m
Wetted surface of hull	S_{hull}	1400 m^2
Casing factor	K_c	1.15
Fore body fullness factor	n_f	2.5
Frontal area of hull	A_F	33.8 m^2
Chord of sail	c_{sail}	13.0 m
Thickness of sail	t_{sail}	2.0 m
Wetted surface of sail	S_{sail}	150 m^2
Chord of aft control surface	c_{acs}	3.5 m
Thickness of aft control surface	t_{acs}	0.3 m
Planform area of aft control surface	$A_{plan_{acs}}$	14.7 m^2
Chord of forward control surface	c_{fcs}	1.8 m
Thickness of forward control surface	t_{fcs}	0.3 m
Planform area of forward control surface	$A_{plan_{fcs}}$	3 m^2

Table 4.5 Components of resistance

Component of resistance	Value (N)
Hull friction (inc roughness)	39,700
Hull form (inc allowance for bow)	6200
Sail friction	5400
Sail form	1700
Total aft control surfaces (ACS)	6000
Total forward control surfaces (FCS)	900
Total	59,900

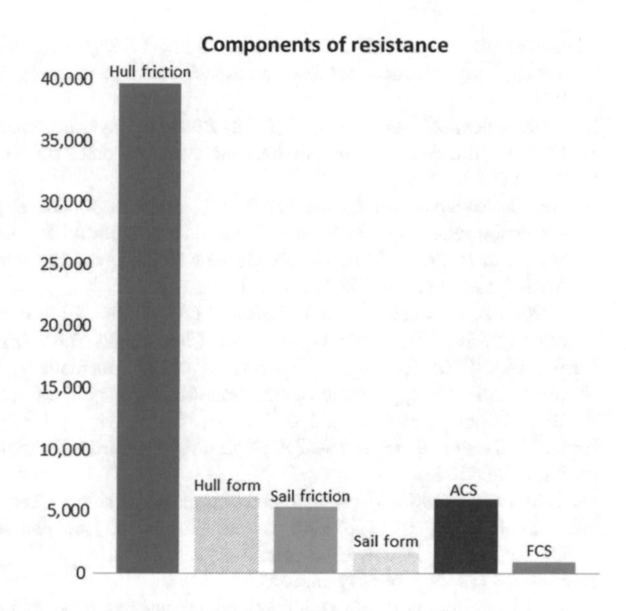

Fig. 4.32 Values of resistance components

As expected, the major component of resistance is the frictional resistance on the hull.

References

Conway AST (2017) Investigation into wakes generated by surface piercing periscopes. Thesis for Doctor of Philosophy, University of Tasmania, May 2017

Conway AST, Renilson MR, Ranmuthugala D, Binns JR (2017a) The effect of speed and geometry on the characteristics of the plume generated by submarine masts. In: Proceedings of warship 2017: naval submarines and UUVs, Royal Institution of Naval Architects, Bath, UK

Conway AST, Ranmuthugala D, Binns JR, Renilson MR (2017b) The effect of geometry on the surface waves generated by vertical surface piercing cylinders with a horizontal velocity. J Eng Marit Environ

Coombs JL, Doolan CJ, Moreau DJ, Zander AC, Brooks LA (2012) Assessment of turbulence models for wing-in-junction flow. In: 18th Australasian fluid mechanics conference, Launceston, Australia, 3–7 Dec 2012

Coombs JL, Doolan CJ, Moreau DJ, Zander AC, Brooks LA (2013) Noise modelling of wing-in-junction flows. In: Acoustics 2013, 17–20 Nov 2013, Victor Harbour, Australia

Crété PA, Leong ZQ, Ranmuthugala D, Renilson MR (2017) The effects of length to diameter ratio on the resistance characteristics for various axisymmetrical hull forms. In: Proceedings of Pacific 2017 international maritime conference, Sydney, Australia, Oct 2017

Dern JC, Quenez JM, Wilson P (2016) Compendium of ship hydrodynamics, practical tools and applications, Les Presses de l'ENSTA, Jan 2016. ISBN-10: 2722509490, ISBN-13: 978-2722509498

Devenport WJ, Agarwal NK, Dewitz MB, Simpson RL, Poddar K (1990) Effects of a fillet on the flow past a wing-body junction. AIAA J 28:2017–2024

Devenport WJ, Simpson RL, Dewitz MB, Agarwal NK (1991) Effects of a strake on the flow past a wing-body junction. In: 29th aerospace sciences meeting, Jan 7–10, 1991/Reno, Nevada, AIAA

Erm, LP, Jones, MB, Henbest SM (2012) Boundary layer trip size selection bodies of revolution. In: Proceedings of the 18th Australasian fluid mechanics conference, Launceston, Australia, 3–7 Dec 2012

Fureby C, Anderson B, Clarke D, Erm L, Henbest S, Giacebello M, Jones D, Nguyen M, Johansson M, Jones M, Kumar C, Lee S-K, Manovski P, Norrison D, Petterson K, Seil G, Woodyatt B, Zhu S (2015) Unsteady flow about a generic submarine—a modelling capability. MAST Asia, Pacifico, Yokohama, Japan

Fu S, Xiao Z, Chen H, Zhang Y, Huang J (2007) Simulation of wing-body junction flows with hybrid RANS/LES methods. Int J Heat Fluid Flow 28(2007):1379–1390

Gertler M (1950) Resistance experiments on a systematic series of streamlined bodies of revolution—for application to the design of high-speed submarines, David W Taylor Model Basin Report C-297, April 1950

Hama FR, Long JD, Hegarty JC (1957) On transition from laminar to turbulent flow. J Appl Phys 28(4):388–394

Harvald Sv AA (1983) Resistance and propulsion of ships. Ocean engineering series. Wiley

Hazarika, BK, Raj RS (1987) An investigation of the flow characteristics in the Blade Endwall Corner Region. NASA Contractor Report 4076

Hoerner SF (1965) Fluid-Dynamic Drag

ITTC (2011a) International towing tank conference recommended procedures and guidelines, ship models, Procedure number: 7.5-01-01-01

ITTC (2011b) International towing tank conference recommended procedures and guidelines, ship models, Procedure number: 7.5-02-02-01

ITTC (2017) Report of resistance committee to the 28th international towing tank conference, Wuxi, China, 2017

Jiménez JM, Smits AJ (2011) Tip and junction vortices generated by the sail of a yawed submarine model at low Reynolds Numbers. J Fluids Eng I33(3):034501-1-4

Jones DA, Clarke, DB (2005) Simulation of a wing-body junction experiment using the fluent code. Defence Science and Technology Organisation, report number: DSTO-TR-1753

Jones MB, Erm LP, Valiyff A, Henbest SM (2013) Skin-friction measurements on a model submarine. Defence Science and Technology Organisation report: DSTO-TR-2898

Leong ZQ, Ranmuthugala D, Renilson MR (2015) Resistance as a function of L/D ratio characteristics for various axisymmetrical hull forms. Australian Maritime College, Tasmania, Australia

Leong ZQ (2017) Personal communication

Liu Z, Xiong Y, Wang Z, Wang S (2010) Numerical simulation and experimental study of the new method of horseshoe vortex control. J Hydrodyn 22(4):572–581

Liu Z, Xiong Y, Tu C (2011) Numerical simulation and control of horseshoe vortex around an appendage-body junction. J Fluids Struct 27(1):23–42

Liu Z, Xiong Y (2014) The method to control the submarine horseshoe vortex by breaking the vortex core. J Hydrodyn 26(4):637–645

Moonesun M, Korol Y (2017) Naval submarine body form design and hydrodynamics. Lambert Academic Publishing. ISBN: 978-620-2-00425-1

Olcmen SM, Simpson RL (2006) Some features of a turbulent wing-body junction vortical flow. Int J Heat Fluid Flow 27(2006):980–993

Overpelt B, Nienhuis B (2014) Bow shape design for increased performance of an SSK submarine. In: Proceedings of warship 2014, Naval Submarines and UUVs, Bath, UK, June 2014

Rawson KJ, Tupper EC (2001) Basic ship theory, 5th edn. Butterworth-Heinemann

Renilson MR, Ranmuthugala D (2012) The effect of proximity to free surface on the optimum length/diameter ratio for a submarine. In: First international conference on submarine technology and marine robotics (STaMR 2012), Chennai, 13–14 Jan 2012

Seil, GJ, Anderson B (2012) A comparison of submarine fin geometry on the performance of a generic submarine. In: Proceedings of Pacific 2012 international maritime conference, Sydney, 2012

Shen YT, Hughes MJ, Hughes JJ (2015) Resistance prediction on submerged axisymmetric bodies fitted with turbulent spot inducers. J Ship Res 59(2):85–98

Simpson RL (2001) Junction flows. Annu Rev Fluid Mech

Stanbrook A (1959) Experimental observation of vortices in wing-body junctions. aeronautical research council reports and memoranda, Ministry of Supply, RAE Report Aero. 2589

Toxopeus SL, Kuin RWJ, Kerkvliet M, Hoeijmakers H, Nienhuis B (2014) Improvement of resistance and wake field of an underwater vehicle by optimising the fin-body junction flow with CFD. In: OMAE ASME 33rd international conference on ocean, offshore and Arctic engineering, San Francisco, CA, 2014

Warren CL, Thomas MW (2000) Submarine hull form optimisation case study. Naval Eng J, pp 27–39

Chapter 5
Propulsion

Abstract The efficiency and acoustic performance of any propulsor will be affected by the flow into it. This is determined by: the hull shape, particularly the aft body and the tail cone angle; the casing; the sail; and the aft appendages. There will be an uneven wake field into the propulsor which will depend on the sail design and aft control surface configuration. This will result in fluctuating forces, causing vibration and noise. The quality of the flow into the propulsor can be assessed quantitatively using either the Distortion Coefficient, or the Wake Objective Function, and these are both explained. Results are presented to estimate the Taylor wake fraction, and the thrust deduction fraction as functions of the tail cone angle and the ratio of propeller diameter to hull diameter. The hull efficiency, which is the ratio of effective power to thrust power, can be estimated. The relative rotative efficiency is the ratio of the open water propulsive efficiency to the efficiency of the propulsor when operating in the wake. The Quasi Propulsive Coefficient (QPC) is the ratio of useful power to the power delivered to the propeller. Submarines are often propelled by a large optimum diameter single propeller. It is important to avoid cavitation, and the Cavitation Inception Speed depends on depth of submergence. Blade number is important, and this is discussed. Many submarines use pumpjets, which comprise two or more blade rows within a duct. The principles of pumpjets are discussed, along with some design guidance. The diameter of a pumpjet is usually smaller than that of a propeller, resulting in a lower rotor tip speed. Contra-rotating propulsion; twin propellers; podded propulsion; and rim driven propulsion are also discussed. Propulsor performance can be assessed using either the thrust identity or torque identity method, and both are described.

5.1 Propulsor/Hull Interaction

5.1.1 Introduction

Modern submarines are normally propelled by a single propulsor on the axis of the submarine as shown in Fig. 5.1.

M. Renilson, *Submarine Hydrodynamics*,
https://doi.org/10.1007/978-3-319-79057-2_5

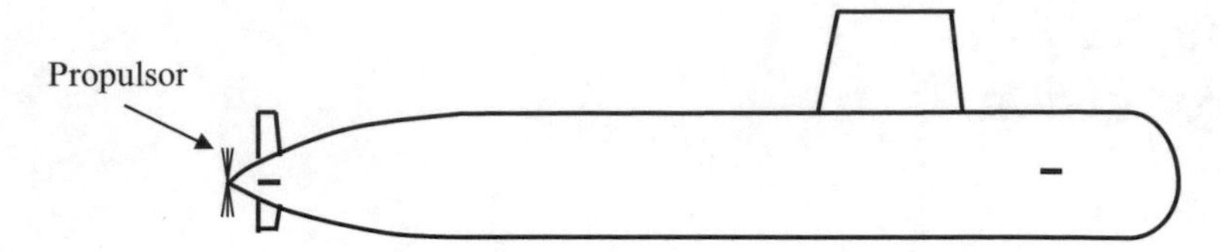

Fig. 5.1 Typical location of propulsor on modern submarine

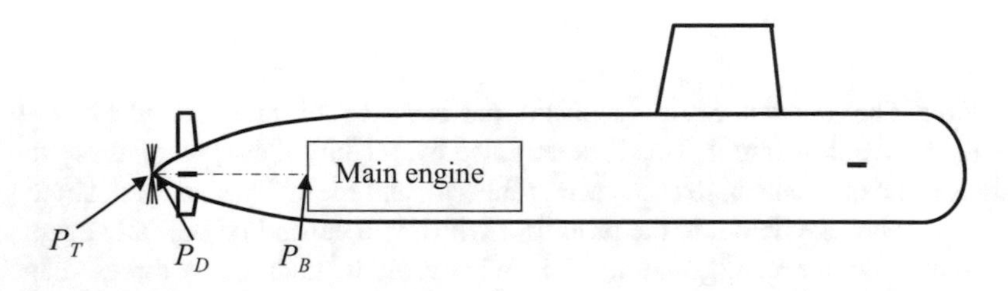

Fig. 5.2 Diagram showing the definitions of power along the shaft line

The effective power, P_E, required to propel a submarine with resistance R, at a velocity V is given by Eq. 5.1.

$$P_E = RV \tag{5.1}$$

The interaction between the hull and the propulsor affects the propulsion of the submarine. This is illustrated in Fig. 5.2.

In Fig. 5.2, P_T is the thrust power from the propulsor, P_D is the power delivered to the propeller, and P_B is the power from the machinery. The thrust power is given by Eq. 5.2.

$$P_T = V_a T \tag{5.2}$$

In Eq. 5.2, V_a is the velocity of advance of the propulsor, and T is the propulsor thrust. V_a is not the same as the velocity of the submarine, due to the wake (sub Sect. 5.1.3), and T is not the same as the resistance of the submarine, due to the thrust deduction caused by the propulsor lowering the pressure over the aft portion of the submarine (sub Sect. 5.1.4). The difference between the thrust power and the effective power is the hull efficiency, η_H, given in Eq. 5.3.

$$\eta_H = \frac{P_E}{P_T} \tag{5.3}$$

The ratio of thrust power, to delivered power is the propulsor efficiency. Because the propulsor is operating behind the submarine this will be different to the efficiency of the propulsor in open water. The ratio of propulsor efficiency for the propulsor operating behind the submarine and the propulsor efficiency in open water is the relative rotative efficiency, η_R, as given in Eq. 5.4. Note that this is not actually an "efficiency" but a ratio of efficiencies.

$$\eta_R = \frac{\eta_B}{\eta_O} \tag{5.4}$$

The Quasi Propulsive Coefficient (QPC) is the ratio of the useful, or effective, power, P_E, to the power delivered to the propeller, P_D, as given in Eq. 5.5.

$$\text{QPC} = \eta_H \eta_R \eta_O \tag{5.5}$$

5.1.2 Flow into the Propulsor

The efficiency and acoustic performance of any propulsor will be significantly affected by the inflow to it. This is determined by: the shape of the hull, particularly the aft body and the tail cone angle; the presence and size of any casing; the shape and size of the sail; and the size and configuration of the aft appendages.

The propulsor will experience different flow conditions when operating behind the submarine as compared to when it is in open water without the presence of the submarine ahead of it. This is caused by the wake of the submarine which has a major effect on the propulsor. In principle the wake is very similar to that for surface ships.

The effect of the wake into the propulsor is illustrated schematically in Fig. 5.3. In this figure a contour of constant velocity is shown. The velocity inboard of the contour line is lower than that on the contour line, and the velocity outboard of it is higher.

The left hand side of the figure shows an X-form shape, with the contour of constant velocity shown as a dashed line. There is a slightly lower velocity at the top, caused by the wake from the sail. There is also a lower velocity in line with the two planes.

The right hand side of the figure shows a cruciform shape, with the contour of constant velocity shown as a full line. There is a considerably lower velocity at the top, caused by the sail and the upper rudder. There is also a lower velocity at the side in the wake of the horizontal stabilizer, and at the bottom, in line with the lower rudder.

The effect that the change in the velocity of the inflow has on the propulsor blade is illustrated in Fig. 5.4. The resultant velocity into the propulsor blade is made up of the axial velocity, V^*, and the circumferential velocity at the blade ($=\pi n D_{local} - V_\theta$), where n is the rotational speed in revolutions per second, D_{local} is the local diameter, and V_θ is the local tangential velocity in the wake.

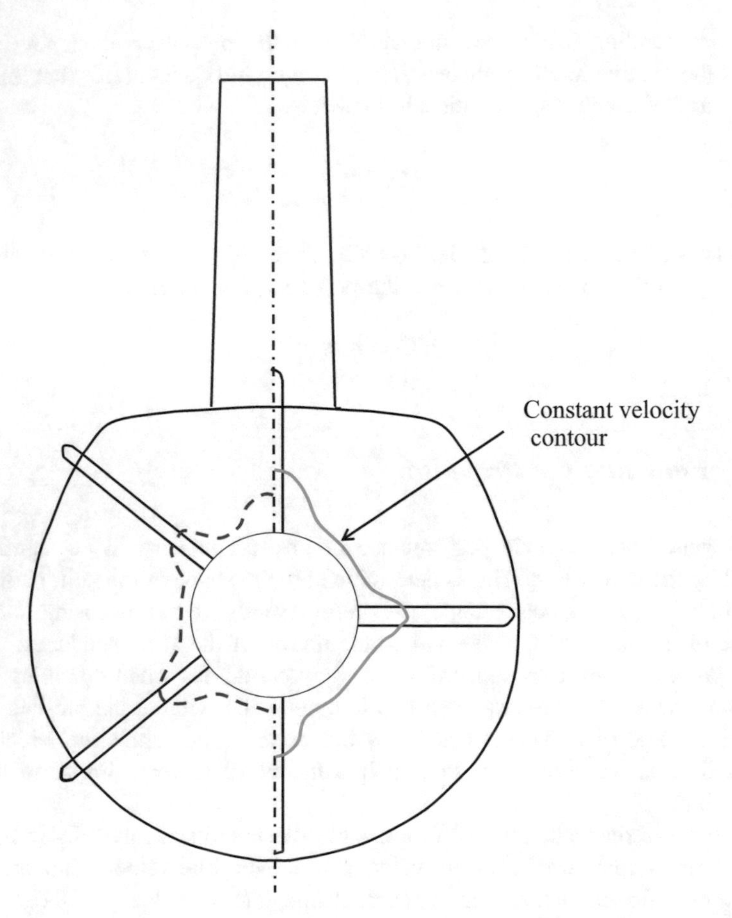

Fig. 5.3 Schematic of wake at the propulsor plane (left hand side illustrates X-form and right hand side illustrates cruciform)

Figure 5.4a illustrates the case when the local axial velocity, V^*, is relatively high. This represents an angular position on the circumference where the wake is small. As can be seen, the angle of the resultant velocity to the blade is small, which is the design condition. Figure 5.4b illustrates the case where the local axial velocity is lower, as it would be at an angular position on the circumference which is in line with an appendage. In this case the angle between the resultant velocity and the blade (the angle of attack) is much larger than in Fig. 5.4a, causing a greater lift and drag on the blade.

Thus, the force on the blade will vary with circumferential position of the blade, resulting in a vibration of the propulsor at a frequency corresponding to rotational speed of the propulsor, the number of blades, and the number of regions of higher wake. This can also be transmitted to the shaft, potentially causing vibration in the

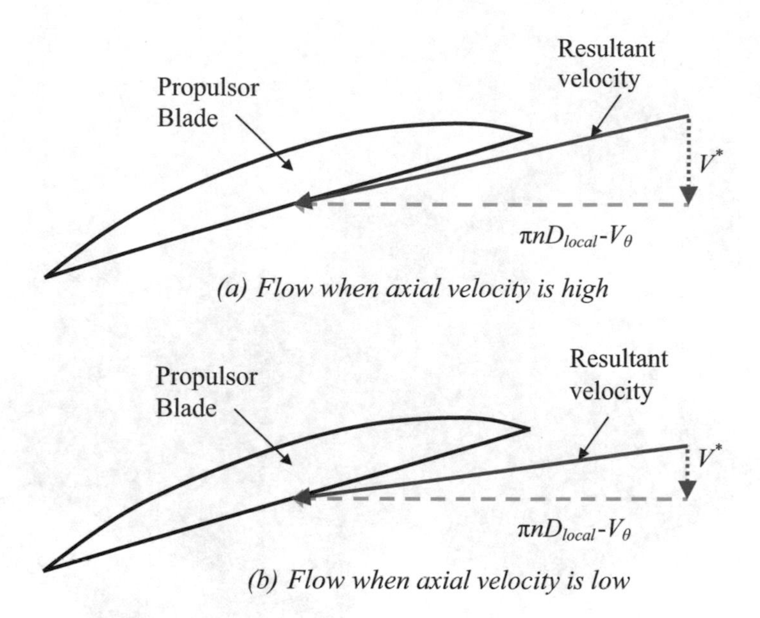

Fig. 5.4 Flow into propulsor blade

whole drive chain, and even in the hull. Care needs to be taken when selecting the number of propulsor blades, related to the circumferential wake pattern.

Because of this it is common practice to determine the wake flow into the propulsor as part of the propulsor design. Generally both unpowered and powered wake surveys are carried out, with both axial and tangential components obtained. This can be done with a physical model in either water or air. Figure 5.5 shows such a set up using Particle Image Velocimetry in a wind tunnel.

Alternatively, good results can generally be obtained using Computational Fluid Dynamics.

The magnitude of the circumferential variation in the wake can be quantified using either the Distortion Coefficient (D_C) proposed by Seil and Anderson (2012) or the Wake Objective Function (WOF) proposed by van der Ploeg (2012, 2015). Both of these were developed to assess the distortion in the wake field to make it possible to compare the effects of the aft body on the propulsor.

The Distortion Coefficient takes into account the circumferential variation of the wake about its mean value at a given radius. For each radius the average axial wake fraction, $\bar{w}$ can be determined by integration around the circle. The Distortion Coefficient is then the standard deviation of the wake fraction about its mean at that radius, and is given by Eq. 5.6, for n measurement points spaced uniformly around the circle.

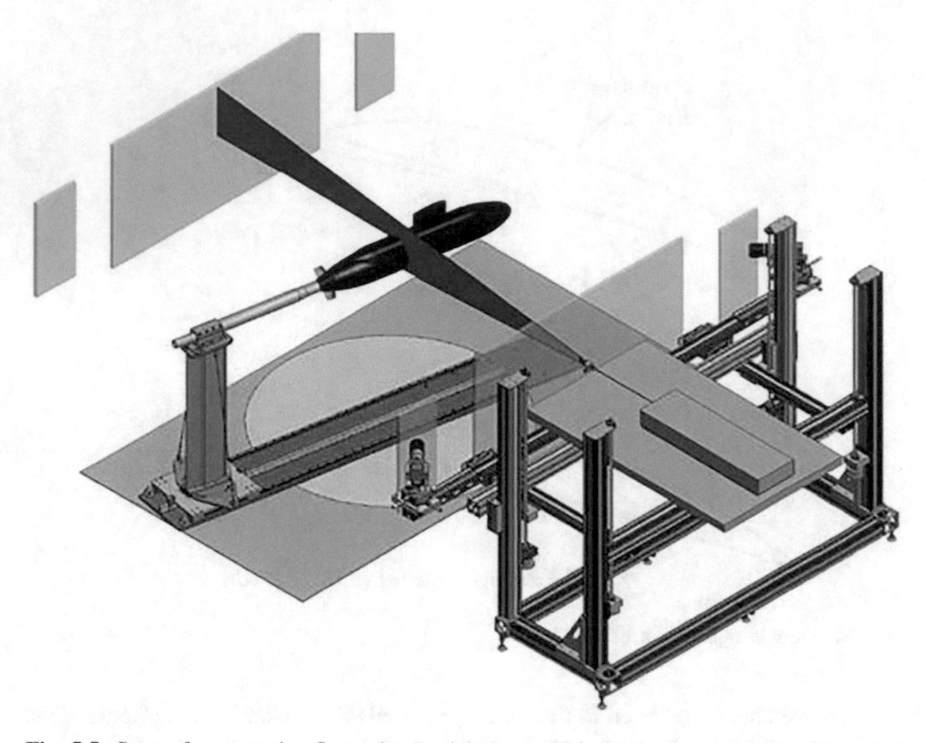

Fig. 5.5 Set up for measuring flow using Particle Image Velocimetry in a wind tunnel (courtesy DST Group)

$$D_C = \sqrt{\frac{\sum_{i=1}^{n} (w_i - \bar{w})^2}{n - 1}} \quad (5.6)$$

In Eq. 5.6 the subscript "*i*" refers to the value at point number *i*. The number of measurement points needs to be sufficiently large that D_C is independent of the number of points used.

The Distortion Coefficient, D_C, can then be plotted as a function of radius, as shown in Fig. 5.6, taken from Seil and Anderson (2012). The effect of the submarine geometry (in this case the sail, or fin, design) on the radial distribution of the Distortion Coefficient can be clearly seen. This makes it possible to quantify the effect of the hull geometry on the flow into the propeller.

An alternative approach is to use a Wake Objective Function (WOF) proposed by van der Ploeg (2012, 2015) which is based on the variation of the local angle of the flow to the propeller blade, β, obtained from Eq. 5.7.

$$\beta = \arctan\left(\frac{V^*}{\pi n D_{local} - V_\theta}\right) \quad (5.7)$$

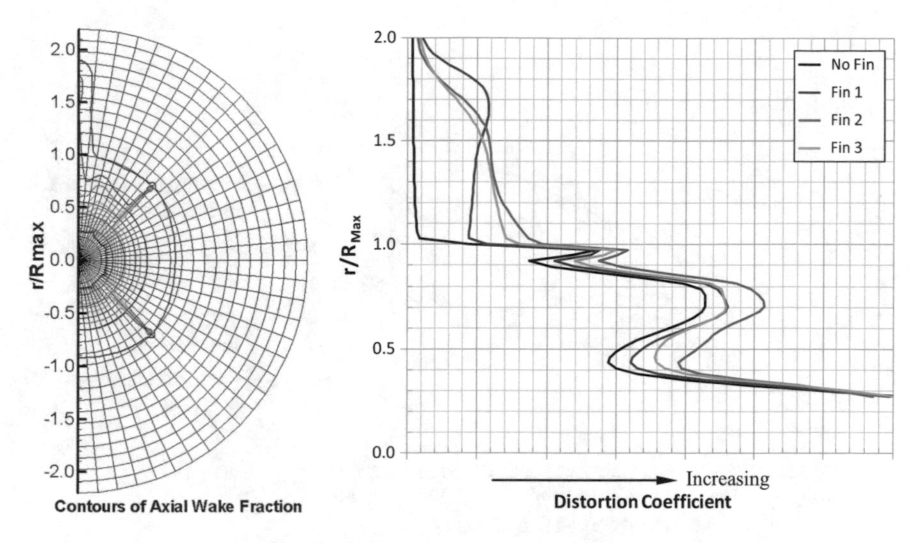

Fig. 5.6 Radial variation of wake Distortion Coefficient (from Seil and Anderson 2012)

The Wake Objective Function is given by Eq. 5.8, where f is a weighting function that van der Ploeg proposed be used to make a particular region of the propeller disc more or less important.

$$\mathrm{WOF} = \frac{\int_{r=hub}^{r=tip} \oint_{\theta} \left|\frac{\partial \beta}{\partial \theta}\right| f(\theta, r) d\theta r d r}{\int_{r=hub}^{r=tip} \oint_{\theta} f(\theta, r) d\theta r d r} \tag{5.8}$$

Note that the WOF obtained from Eq. 5.8 is a single value for the whole propeller disc. If the weighting function is not used, the Wake Objective Function can be calculated as a function of radius using Eq. 5.9. This can be plotted as a function of radius in a similar manner to the Distortion Coefficient.

$$\mathrm{WOF}(r) = \oint_{\theta} \left|\frac{\partial \beta}{\partial \theta}\right| d\theta \tag{5.9}$$

5.1.3 Wake

The wake fraction, w, is defined by Eq. 5.10.

$$w = \frac{V - V_a}{V} \tag{5.10}$$

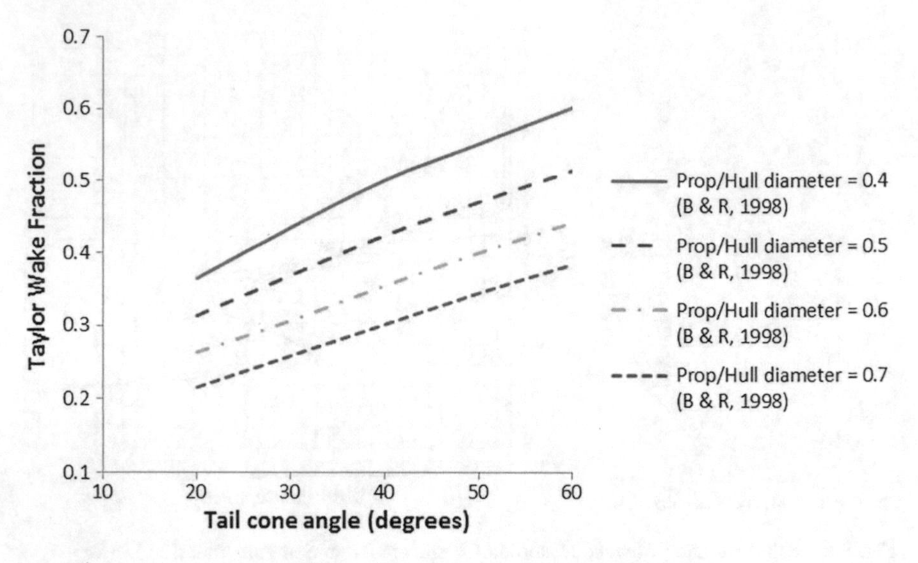

Fig. 5.7 Taylor wake fraction (adapted from Burcher and Rydill 1998)

In this equation, w, is known as the Taylor wake fraction, V, is the vessel velocity, and V_a is the velocity of advance of the propulsor.

The Taylor wake fraction will depend on the tail cone angle, and the ratio of propeller diameter to hull diameter, as shown in Fig. 5.7 adapted from Burcher and Riddell (1998). Note that the total tail cone angle (twice the half tail cone angle) is used in this figure.

5.1.4 Thrust Deduction

The other major effect that the propulsor will have is to generate a "pull" on the aft end of the submarine, due to the low pressures ahead of it. This is accounted for by assuming a deduction in thrust. The thrust deduction fraction, t, is given by Eq. 5.11.

$$t = \frac{T - R_T}{T} \tag{5.11}$$

In this equation, t, is the thrust deduction fraction, T, is the propulsor thrust, and R_T is the total hull resistance.

The thrust deduction fraction will depend on the tail cone angle, and the ratio of propeller diameter to hull diameter, as shown in Fig. 5.8, adapted from Burcher and Riddell (1998) and Kormilitsin and Khalizev (2001). Note that the total tail cone angle (twice the half tail cone angle) is used in this figure.

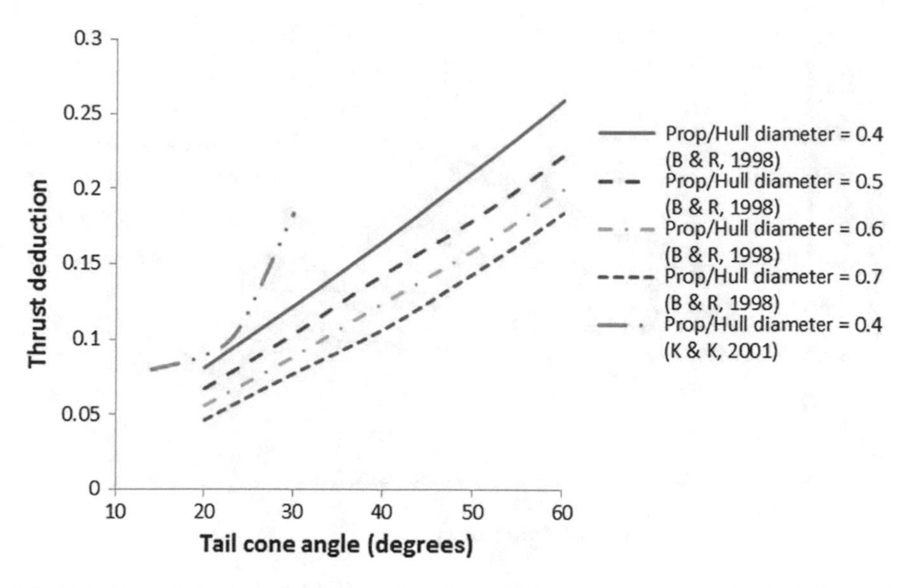

Fig. 5.8 Thrust deduction (adapted from Burcher and Rydill 1998, and Kormilitsin and Khalizev 2001)

5.1.5 Hull Efficiency

The hull efficiency, η_H, is the ratio of the effective power to the thrust power. It is important to note that the hull efficiency is not efficiency as such, but the ratio of efficiencies. Thus, it is possible for the hull efficiency to be greater than one. Hull efficiency is defined in Eq. 5.3. When Eqs. 5.1 and 5.2 are substituted into Eq. 5.3 then this become Eq. 5.12.

$$\eta_H = \frac{P_E}{P_T} = \frac{(1-t)}{(1-w)} \tag{5.12}$$

P_E is the effective power; P_T is the thrust power; t is the thrust deduction; and w is the Taylor wake fraction.

Using the data in Figs. 5.7 and 5.8 the hull efficiency obtained is given in Fig. 5.9. Note that the total tail cone angle (twice the half tail cone angle) is used in this figure. As can be seen, the hull efficiency is always greater than one. This is due to the high wake, compared to the thrust deduction fraction. At an early stage of the design this figure can be used to estimate the hull efficiency, however self-propelled model experiments would usually be used to refine this at a later stage in the design.

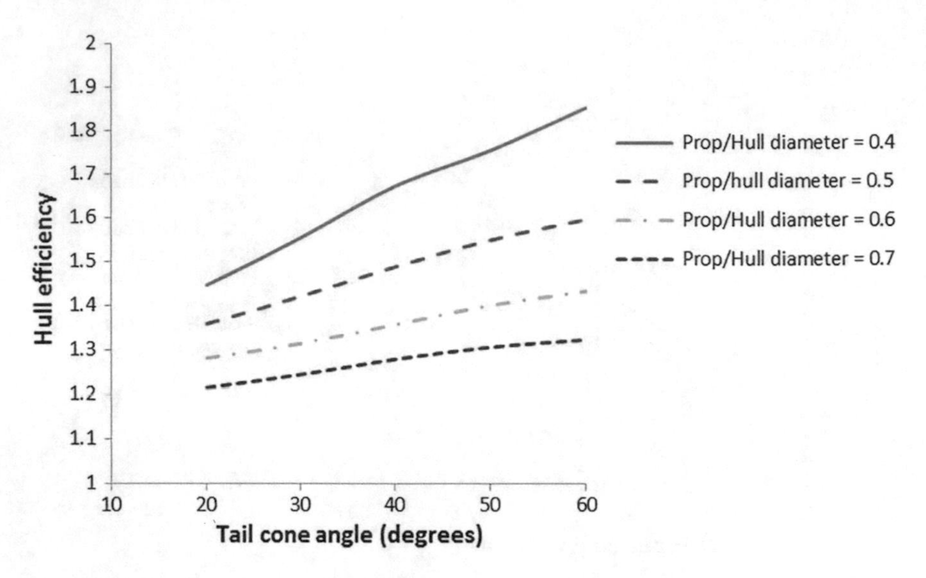

Fig. 5.9 Hull efficiency (using data from Burcher and Rydill 1998)

5.1.6 Relative Rotative Efficiency

The relative rotative efficiency, η_R, is the ratio of the open water propulsive efficiency, η_O, to the efficiency of the propulsor when operating in the wake behind the submarine. For a submarine with a single propulsor on the axis the value of η_R will depend on the shape of the aft body (characterized by the tail cone angle) and appendages. It is also dependent on the ratio of the propeller diameter to hull diameter.

For values of the ratio of propeller diameter to hull diameter in the range of 0.4–0.7, and full tail cone angles of 20–50°, η_R will be approximately 1.05, indicating that the propeller is more efficient when operating behind the submarine, than in open water. Part of the reason for this is the flow straightening effects of the appendages, which will reduce the losses due to swirl.

For small propeller diameters, and large tail cone angles, the value of η_R can reduce to below unity.

On the other hand, in the extreme case where the propeller diameter is much greater than the hull diameter, and the tail cone angle is small, then the value of η_R must tend to unity, as the propeller will be operating close to its open water condition.

5.1.7 Quasi Propulsive Coefficient

The Quasi Propulsive Coefficient (QPC) is the ratio of useful power to the power delivered to the propeller. It is made up of all the various efficiencies, as given in Eq. 5.13.

$$\mathrm{QPC} = \eta_H \eta_O \eta_R \tag{5.13}$$

where: η_H is the hull efficiency; η_O is the open water propeller efficiency; and η_R is the relative rotative efficiency. Generally for submarines with a single propulsor on the axis the value of QPC is between 0.8 and 1.0.

5.2 Axisymmetric Hull with Single Propeller

The majority of modern conventional submarines are propelled by a single propeller on the axis, as shown in Fig. 5.10. This arrangement makes it possible to use an optimum large diameter propeller, with low rpm, resulting in a highly efficient propeller.

Fig. 5.10 Display model showing single propeller configuration

The preliminary design of the propeller is very similar to that used for a surface ship, as discussed extensively by a number of text books in the field, including Carlton (2007). Use can be made of systematic series data, as discussed in van Lammeren et al. (1969).

However, one fundamental difference between the design of a propeller for a submarine, and that for a surface ship, is that because the submarine does not generate wavemaking resistance, the resistance is almost proportional to velocity squared. The result of this is that regardless of speed through the water the advance coefficient, J, that the propeller operates at is almost constant.

The principal dimensions of a propeller are:

(a) number of blades;
(b) diameter;
(c) pitch;
(d) rotational speed; and
(e) blade area.

The first requirement for a submarine propeller design is the selection of the blade number. This is based on the need to minimise the acoustic noise signature generated by fluctuating forces on the propeller caused by it operating in the wake behind the submarine, as discussed in sub Sect. 5.1.2.

As can be seen in the right hand side of Fig. 5.3, for a cruciform stern there are typically four areas of reduced flow around the circumference, due to the presence of the planes and the sail/casing. Hence, a four bladed propeller would experience each blade suffering from the reduced inflow at the same time, causing considerable vibration and resulting hydro-acoustic noise. Thus, four blades, or a blade number which is a multiple of four, should be avoided.

With the X-form stern shown on the left hand side of Fig. 5.3 there are five such areas of reduced flow due to the wake, and hence five blades, or a multiple of them, should be avoided.

Ideally, as high a blade number as possible should be selected, which should be a prime number, to avoid possible harmonics. However, there are practical concerns with very high blade numbers, and so consequently it is common practice to use a seven bladed propeller.

Next the diameter, pitch and rotational speed are selected.

For a submarine with a single propeller on its axis the propeller design is not constrained by its diameter, making it possible to select the optimum diameter for the required thrust at a given rotational speed from an existing propeller series. The pitch can then be determined from such series data. This process is the same as for surface ships, and is discussed in a number of text books on the subject, including Carlton (2007).

Although the hub size is generally a lot larger for a submarine than for a surface ship, a propeller series developed for surface ships will generally give a satisfactory preliminary propeller design, and a good estimate of its efficiency, adequate for concept design studies etc. This is because the velocity over the propeller at the hub is much lower than that over the rest of the blade, so its contribution to the overall thrust and torque is generally small. However, if series data for a submarine propeller is available it would clearly be preferable to make use of that.

The required blade area is then selected to avoid cavitation. Too small a blade area will result in too great a load on the blade, and hence a poor cavitation performance.

When a submarine is running deep cavitation should not be a problem, as the substantial head will increase the pressure, and hence cavitation is very unlikely. However, when running close to the water surface, cavitation could be a problem for a propeller with insufficient blade area. Also, when operating in off design conditions, such as accelerating, turning, or braking cavitation may become an issue. Generally, avoiding cavitation in operating conditions for an undamaged submarine propeller is not difficult.

Even when a large blade number is selected, each blade will be operating in the wake, and can suffer a change in loading due to the circumferential change in wake. This can cause vibration and hydro-acoustic noise. Thus, to attempt to smooth this out over each blade, a large amount of skew is generally adopted, as shown in the display model in Fig. 5.10. This means that only a section of the blade will be experiencing the low axial inflow at any one time, and hence the overall effect on the blade will be reduced. However, care needs to be taken with excessive skew to avoid unwanted flexure in the blades, and also large stresses in the blades when applying astern thrust.

Once the principal dimensions of the propeller have been selected it is possible to consider the design features which will further reduce acoustic noise signature. For example, suction and pressure-side cavitation can normally be avoided by a thicker section profile than used for a comparable propeller for a surface ship. In order to avoid tip-vortex and hub-vortex cavitation submarine propellers usually have a reduction in loading at their tips and hubs. In addition, the trailing edge region needs careful design to reduce trailing edge noise. It is not sufficient simply to reduce singing, but the whole noise spectrum needs to be considered. Hence submarine propellers are not normally provided with anti-singing trailing edges (Anderson et al. 2009).

An example of a submarine propeller, from *HMS Trafalgar*, is shown in Fig. 5.11.

5.3 Axisymmetric Hull with Single Pumpjet

Pumpjets comprise two or more blade rows within a duct, which can either be rotating (rotor) or static (stator) blades. One of the design objectives is to use the stators to remove the rotational flow imparted by the rotors. This rotational flow, which is always present behind single propellers, represents hydrodynamic loss, as

Fig. 5.11 Propeller from HMS Trafalgar

Fig. 5.12 Pumpjet on submarine model in the cavitation tunnel at SSPA (photo by Sven Wessling, courtesy of SSPA, taken from: SSPA 1993)

energy has been utilised to rotate the flow, which is not used to propel the vessel and this reduces the efficiency. A photograph of a pumpjet in the cavitation tunnel at SSPA is given in Fig. 5.12, taken from SSPA (1993), with permission.

The primary motivation for the design of pumpjets for application to submarines is to reduce the hydro-acoustic signature, however it is also claimed that it can have a higher efficiency (Vinton et al. 2005).

The pumpjet provides a means of controlling the flow velocities over the blades and enables a direct trade-off of cavitation performance with efficiency to be achieved. In addition, the duct acts as an end wall for the blades, therefore enabling them to be loaded out to their tips, resulting in a significant reduction in the propulsor diameter compared to an open propeller. This will result in a lower tip speed. However, the gap between the tip of the rotor blade and the duct becomes very important, as this is where cavitation could occur. Ideally the gap should be as small as possible.

A pumpjet can be designed with the stator row aft of the rotor row (post-swirl) or with the stator row ahead of the rotor row (pre-swirl) as shown in Fig. 5.13. A post-swirl pumpjet requires additional struts forward of the rotor to support the duct, as can be seen in Fig. 5.13a.

The stators in post-swirl pumpjets can contribute of the order of 25% of the total propulsor thrust, reducing the required loading on the rotors, and hence their tendency to cavitate (Clarke 1988). In addition, the duct can be designed to decelerate the flow, hence increasing the pressure. Both these measures will help to control cavitation—potentially an important issue for a submarine in the high speed regime.

On the other hand, the stators in a pre-swirl pumpjet contribute a drag which means that the actual thrust on the rotor will be greater than the total thrust on the pumpjet, reducing its cavitation performance. Because of the drag on the stators, pre-swirl pumpjets are generally less efficient than post-swirl ones, although their efficiency is likely to be as good as, or better than, a propeller.

In a post-swirl pumpjet the rotors are operating directly in the wake from the appendages (sail and aft planes) resulting in the generation of narrowband radiated noise at blade rate frequencies. This can be significant (Clarke 1988).

With a pre-swirl pumpjet the stators filter out the wakes from the appendages prior to them reaching the rotor, resulting in a quieter propulsor (Clarke 1988).

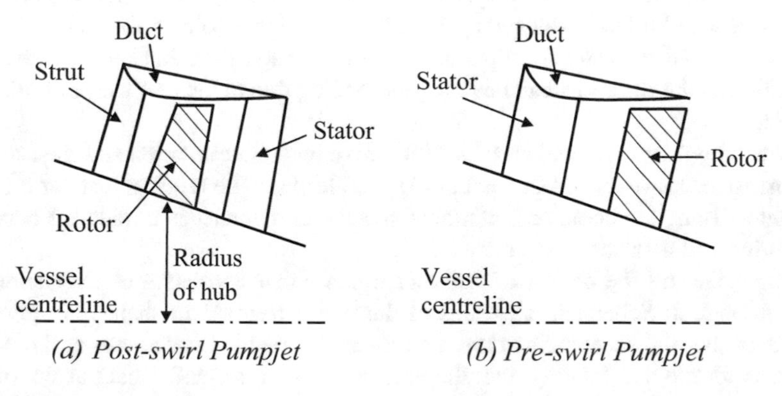

Fig. 5.13 Schematic of post-swirl and pre-swirl pumpjets (flow direction is from left to right)

Also, the flow velocity over the stators will be less for a pre-swirl pumpjet compared to a post-swirl pumpjet. Thus, if cavitation is not an issue due to the depth the boat is operating at, a pre-swirl pumpjet is likely to be quieter than a post-swirl one.

Pumpjets can also be designed with two stator rows—one ahead of, and one behind, the rotor.

In principle, pumpjets could also have more than one row of rotors and/or could also use contra-rotating rotors to remove the rotational flow, along the same principle as contra-rotating propellers used for torpedoes and surface ships.

Pumpjets generally have a large number of blades, both in the rotor and stator rows. The number of blades is likely to be a prime number, to avoid harmonics. It is important to avoid having the same number of blades in more than one row.

As the number of blades has a key influence on the performance of the pumpjet, in particular the acoustic performance, the number of blades on any boat in service is very highly classified.

Despite the lower diameter, a pumpjet is usually much heavier than the equivalent propeller as there are many more parts, including the duct and the stators, compared to a propeller.

The blades on the pumpjet are normally individually manufactured and attached to the hub. This results in a more complex hub than would be required for a fixed pitch propeller, meaning that the hub is likely to require a larger diameter. Hence, a pumpjet is likely to be mounted further forward on the submarine, where the hull diameter is greater. In addition, the larger mass of the pumpjet can cause a problem if placed too far aft, which is another reason for moving it forward compared to the location of a conventional propeller.

The primary parameter influencing cavitation performance is the blade area. Too small a blade area will result in too great a load on the blade, and hence a poor cavitation performance. If the diameter is small it will be difficult to provide sufficient blade area, and so the required blade area has a major influence on the pumpjet diameter necessary for cavitation performance.

A further important parameter is the rotor tip speed. Higher tip speeds will result in a higher possibility of tip-vortex cavitation, which is usually the type of cavitation which affects Cavitation Inception Speed (CIS). For a given propulsor rotational speed a larger diameter will result in a higher rotor tip speed.

The required diameter for a pumpjet will be smaller than that for an equivalent propeller, as the rotor tips can have higher loading due to the end wall provided by the duct.

Low propulsor rotational speed will improve both the cavitation and the acoustic performance, however, lower rotational speed leads to the requirement for a larger diameter. Thus, the choice of rotational speed and diameter is a trade-off between cavitation and acoustic performance.

Depending on the design of the duct it can either accelerate or decelerate the flow through it. Schematic diagrams of the two extreme duct shapes are given in Fig. 5.14. For the accelerating duct the area at the inlet is greater than at the outlet, whereas for the decelerating duct the area at the inlet is smaller than at the outlet.

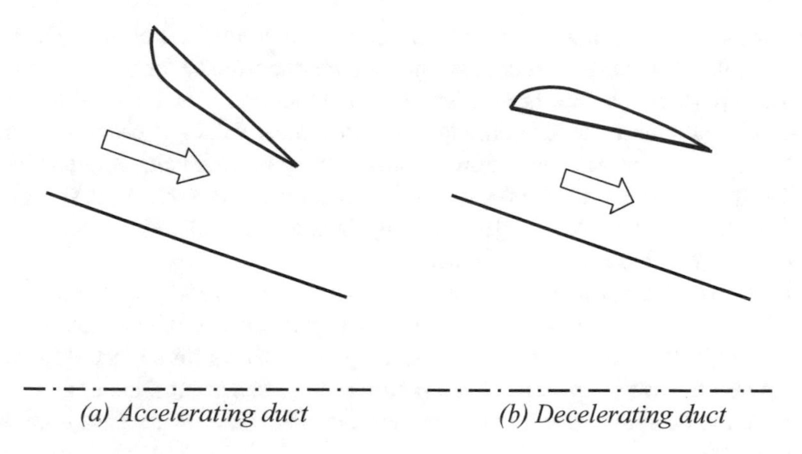

Fig. 5.14 Schematic of accelerating and decelerating ducts (flow direction is from left to right)

For designs with accelerating ducts, the duct can actually contribute to a positive thrust, increasing the efficiency of the rotor/duct combination. This concept is often used for surface ships requiring high thrust at low speed, such as those engaged in towing operations. On the other hand, vessels with lower thrust and higher speed—hence lower propeller loading—do not normally benefit from an accelerating duct.

The higher water velocity over the blades when fitted with an accelerating duct will result in a lower pressure, and hence reduced cavitation performance.

Decelerating ducts reduce the water velocity over the blades, increasing the pressure, and hence can result in better cavitation performance. Against this, these ducts increase drag, resulting in the need for greater thrust from the blades, and decreasing efficiency.

The optimum duct shape for a submarine will depend on the propulsor loading coefficient, B_p, which is defined in Eq. 5.14.

$$B_p = \frac{NP_S^{0.5}}{V_a^{2.5}} \tag{5.14}$$

where N is the revolutions per minute, P_S is the power delivered to the shaft in horsepower, and V_a is the velocity of advance of the propulsor in knots. As a rule of thumb for surface ships, a duct can be beneficial for B_p values greater than around 40 (Carlton 2007). Submarines generally have B_p values far smaller than this, implying that the duct itself would not be an advantage from a hydrodynamic point of view.

Both stator blades and rotor blades have aspect ratios much smaller than conventional open propellers. They do not have anything like the same level of skew that conventional submarine propellers have.

A well designed pumpjet will result in no rotational flow aft of the unit. As the rotational flow represents lost energy, there is the opportunity for pumpjets to have a higher efficiency than open propellers. On the other hand, the duct and the stators (pre-swirl) can contribute to additional drag for typical values of propulsor loading associated with submarine operations, reducing efficiency. A value for efficiency of 0.833 is quoted for the design (post-swirl) in open water discussed by McCormick and Eisenhuth (1963). A pumpjet operating in the wake behind an axisymmetric body is likely to have a higher efficiency.

The astern performance of a pumpjet is much worse than a conventional open propeller. Part of the reason for the poor astern performance with a pumpjet is the duct design, and in particular the sharp trailing edge. Ducts for surface ships which are required to have good astern thrust performance, such as tugs, are designed with more rounded trailing edges, which reduce the efficiency in the ahead condition (Carlton 2007).

5.4 Other Configurations

5.4.1 Contra-rotating Propulsion

Contra-rotating propulsion consists of two propellers rotating in opposite directions on the same shaft. The aft propeller recovers some of the rotational energy imparted by the forward propeller. Thus, in principle, the propulsive efficiency can be much greater than for a single propeller.

Contra-rotating propulsion was used successfully on the USS *Jack* from 1967 to 1989, demonstrating a 10% increase in propulsive efficiency (Dutton 1994).

As contra-rotating propulsion spreads the load over two propellers the blade loading is decreased, and hence cavitation is reduced. In addition, the propulsor rotational speed and/or the diameter can be reduced, again improving cavitation performance. However, not so much is known about the non-cavitating acoustic performance of contra-rotating propulsion, and it is not commonly used for submarines.

On the other hand, contra-rotating propulsion has been common in the past for torpedo propulsion, as shown in Fig. 5.15. This is partly due to the reduction in the rolling moment that would be caused by a single propeller. However, modern torpedoes tend to use pumpjet propulsion, discussed in Sect. 5.3.

Fig. 5.15 Contra-rotating propellers on a MK 44 Torpedo

Fig. 5.16 Configuration for twin screw arrangement

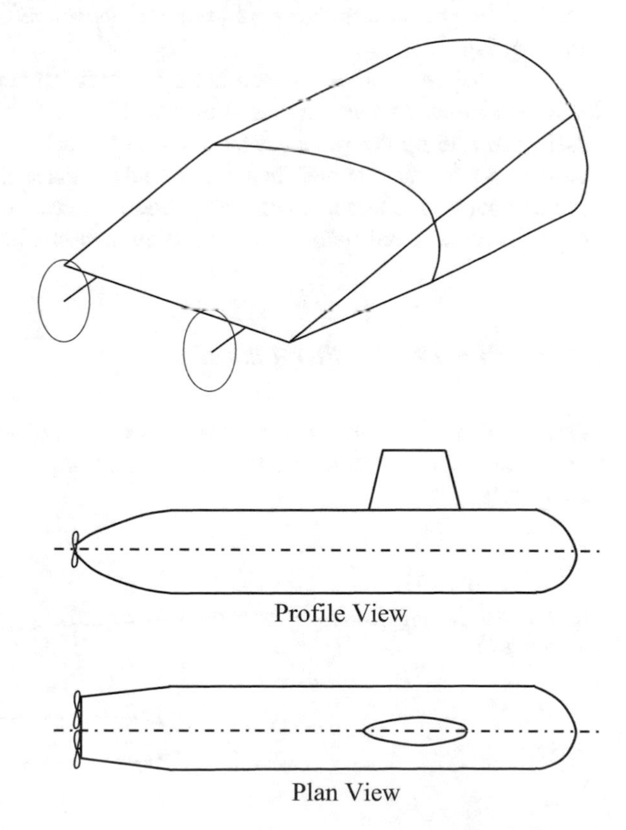

5.4.2 Twin Propellers

In some cases it is desirable to use twin propellers, which also improves the redundancy of the propulsion system. Examples of submarines which use twin screw propellers are the Russian Delta and Typhoon classes.

Changing from a single shaft to a twin shaft power plant of the same total output will result in an increase in nominal displacement of 10–20% (Kormilitsin and Khalizev 2001).

As with all propulsion types, the inflow to the propellers is vital to their acoustic performance. Thus, when twin propellers are being used on an axisymmetric submarine the inflow can be improved by flattening the after body, to result in a shape somewhat like that shown schematically in Fig. 5.16.

5.4.3 Podded Propulsion

Podded propulsion is used for a number of surface ships. It normally comprises an azimuthable pod enclosing an electric motor. Generally, the propeller is arranged ahead of the pod in a tractor configuration, and therefore in undisturbed flow, as shown in Fig. 5.17.

Such a system could be adapted in the future for use on a submarine, particularly for a twin screw arrangement as shown in Fig. 5.18. Note that the pods make it possible to line up the propeller axis with the local flow direction. In addition, if these pods are azimuthable, then these could replace the stern planes. However, various issues such as the electro-magnetic signature and shock resistance would need to be considered before being used on a submarine.

5.4.4 Rim Driven Propulsion

With rim driven propulsion the rotor in a duct is driven through its tip using a permanent-magnetic electric motor, rather than using a shaft arrangement, as shown in Fig. 5.19.

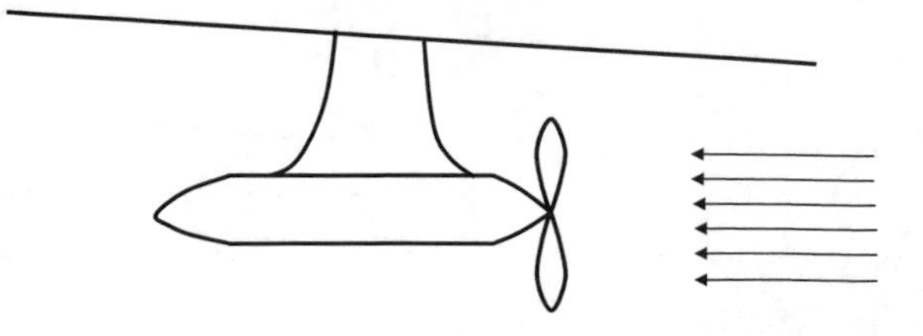

Fig. 5.17 Podded propulsion for a surface ship

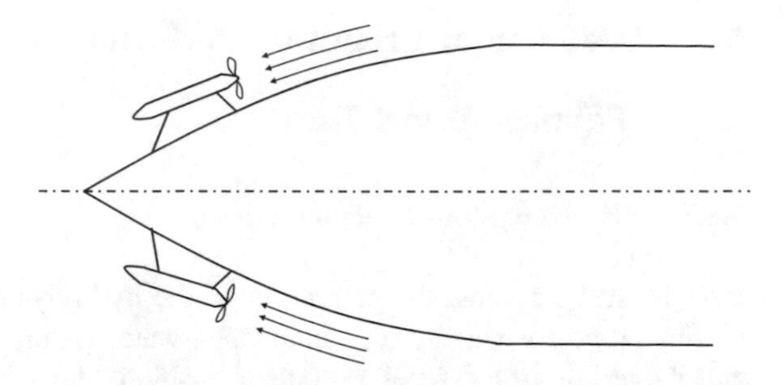

Fig. 5.18 Plan view of possible podded propulsion layout

Fig. 5.19 Typical rim driven propulsor (for a transverse thruster on a surface ship)

The lack of a shaft improves the flow into the blades and hence can reduce vibration, and subsequent hydro-acoustic noise. In addition there is no gap between the outer edge of the rotor and the duct. This results in improved cavitation performance, compared to a shaft driven design, which must incorporate this gap. The inner portion of the rotor blades, which has the lowest velocity, is where the "tip" is, so the resulting vortices will be smaller in magnitude.

As the rotor is driven independently by its outer edge, contra-rotating rotors do not require the mechanical issues associated with a shaft, as discussed in sub Sect. 5.4.1. Also, the rotational speed of the different rotors can be independently controlled.

Finally, since the rotor is driven by a permanent-electric motor in the duct, it can be configured in a similar manner to a podded propulsor, and can be fully azimuthable, if required. Alternatively it can simply replace the single propulsor on the axis of an axisymmetrical hull shape.

5.5 Prediction of Propulsor Performance

5.5.1 Physical Model Tests

5.5.1.1 Hydrodynamic Performance

As with surface ships, the prediction of the hydrodynamic performance of a submarine propulsor will involve both open-water testing of a large scale propulsor, and the testing of a "stock" propulsor "behind" the submarine hull.

The former is often carried out in a cavitation tunnel. Cavitation must be completely avoided on submarine propulsors, but as they usually operate at high pressure due to the large head of water, cavitation inception is not generally such an issue as with surface warships. Equally, it is not important to understand the behaviour of a submarine propulsor under cavitating conditions—as cavitation is completely avoided.

Tests to determine propulsor-hull interaction are conducted in a similar manner to those for a surface ship. The model is usually tested in a towing tank and is inverted using a two strut arrangement as shown in Fig. 4.26. Note that unlike surface ship tests Froude number is not important, so it is possible to conduct the tests at a single speed.

Tests are conducted at various different propulsor rpm, and the propulsor torque and thrust, and the hull resistance are measured. From this, the effective wake and the thrust deduction values at the full scale self-propulsion speed can be obtained, in a similar manner to that used for surface ships. See, for example, ITTC (2014).

The procedure requires prior knowledge of the open-water propeller characteristics for the "stock" propeller being used, and the resistance of the unpropelled submarine.

It is necessary to obtain the full scale value of the resistance from the model test results. One way of doing this is to use the ITTC resistance test procedure (ITTC 2011a), however it must be borne in mind that this has been developed for surface ships, and hence the approach in this procedure to form factors is not applicable. See sub Sect. 4.9.1 for more details.

Once the full scale resistance has been determined the skin friction correction force, F_{D}, to be applied to the model such that the propeller is operating at its full scale self-propulsion point is obtained, as discussed in ITTC (2011b).

Model tests are then conducted with an operating propeller, and the thrust, T_M, and torque, Q_M, on the propeller, and the force on the model in the X direction are measured. The propeller rotational speed is adjusted until the force on the model in the longitudinal direction equals the skin friction correction force, F_{D}. This means that the model is operating at the full scale self-propulsion point.

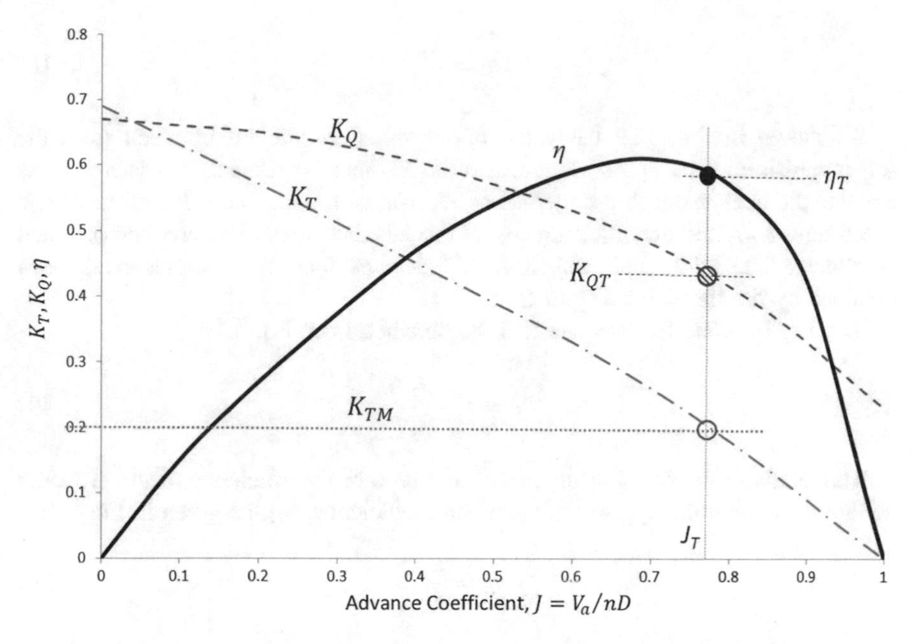

Fig. 5.20 Schematic of open water propeller curve showing thrust identity approach

The thrust deduction fraction, t, can then be obtained from Eq. 5.15.

$$t = \frac{T_M + F_D - R_T}{T_M} \tag{5.15}$$

where R_T is the resistance at model scale.

The measured thrust, T_M, and torque, Q_M, are non-dimensionalised using Eqs. 5.16 and 5.17.

$$K_{TM} = \frac{T_M}{\rho D^4 n^2} \tag{5.16}$$

$$K_{QM} = \frac{Q_M}{\rho D^5 n^2} \tag{5.17}$$

The analysis can then be conducted based on either the "thrust identity" or the "torque identity" approaches. Note that the thrust identity method is recommended by the ITTC.

Thrust identity

For the thrust identity method the thrust coefficient obtained from Eq. 5.16 with the thrust measured in the self-propulsion test is used to determine the equivalent advance coefficient, J_T, as given by Eq. 5.18 and as illustrated in Fig. 5.20.

$$J_T = \frac{V_a}{nD} \tag{5.18}$$

As shown in Fig. 5.20 the value of the thrust coefficient obtained from the self-propulsion test, K_{TM}, is plotted, and the advance coefficient, J_T, where this is equal to the open water thrust coefficient, K_T, can be found. The value of the torque coefficient, K_{QT}, and the efficiency, η_T, at this advance coefficient are then obtained as shown. The additional subscript: "*T*" denotes that these values have been obtained by the thrust identity method.

The Taylor wake fraction can then be obtained from Eq. 5.19.

$$w_T = 1 - \frac{J_T D n}{V} \tag{5.19}$$

The relative rotative efficiency, η_R, is the ratio of the efficiency of the propeller behind the submarine, η_B, to the open water efficiency, η_O, as given in Eq. 5.20.

$$\eta_R = \frac{\eta_B}{\eta_O} \tag{5.20}$$

At the advance coefficient J_T the efficiency of the propeller behind the submarine is given by Eq. 5.21, and the efficiency of the propeller in open water is given by Eq. 5.22.

$$\eta_B = \frac{J_T}{2\pi} \frac{K_{TM}}{K_{QM}} \tag{5.21}$$

$$\eta_O = \frac{J_T}{2\pi} \frac{K_T}{K_Q} \tag{5.22}$$

As can be seen from Fig. 5.19, by definition in this case: $K_{TM} = K_T$ (thrust identity), and $K_Q = K_{QT}$. Thus, the relative rotative efficiency obtained using the thrust identity method is given by Eq. 5.23.

$$\eta_{RT} = \frac{K_{QT}}{K_{QM}} \tag{5.23}$$

Equation 5.23 is the ratio of the torque coefficient in open water at the same advance coefficient, to the torque coefficient measured in the self-propulsion test.

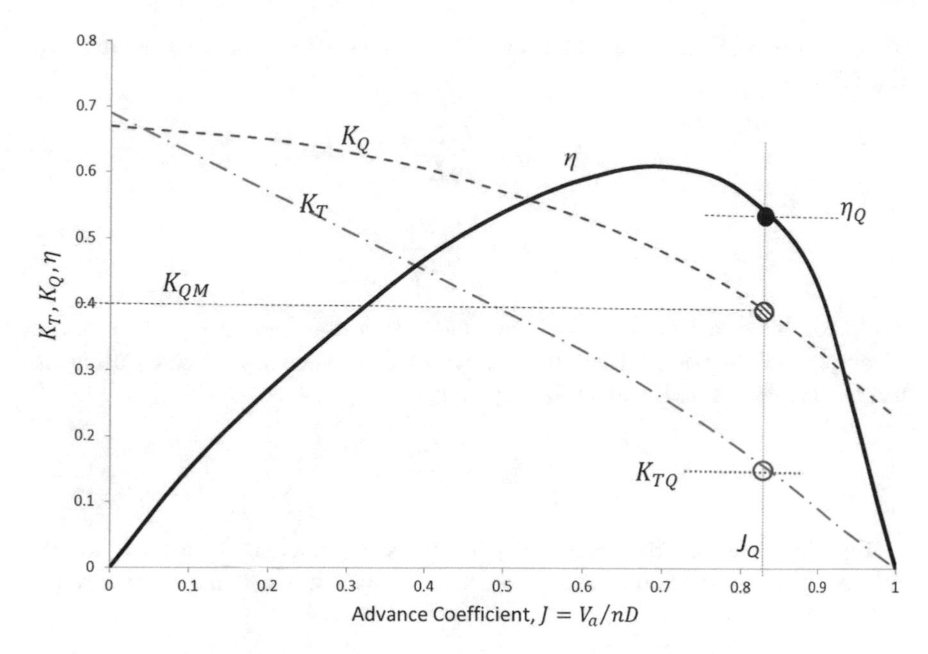

Fig. 5.21 Schematic of open water propeller curve showing torque identity approach

Torque identity

For the torque identity method the torque coefficient obtained from Eq. 5.17 with the torque measured in the self-propulsion test is used to determine the equivalent advance coefficient, J_Q, as given by Eq. 5.24.

$$J_Q = \frac{V_a}{nD} \tag{5.24}$$

This is illustrated in Fig. 5.21. As shown in the figure, the value of the torque coefficient obtained from the self-propulsion test, K_{QM}, is plotted, and the advance coefficient, J_Q, where this is equal to the open water torque coefficient, K_Q, can be found. The value of the thrust coefficient, K_{TQ}, and the efficiency, η_O, at this advance coefficient are then obtained as shown. The additional subscript: "Q" denotes that these values have been obtained by the torque identity method.

The Taylor wake fraction can then be obtained from Eq. 5.25.

$$w_Q = 1 - \frac{J_Q Dn}{V} \tag{5.25}$$

As for the thrust identity method, the relative rotative efficiency is the ratio of the efficiency of the propeller behind the submarine, η_B, to the open water efficiency, η_O. In this case, the comparison is made at the advance coefficient obtained from the torque identity, J_Q. Thus, the efficiency of the propeller behind the submarine is given by Eq. 5.26, and the efficiency of the propeller in open water is given by Eq. 5.27.

$$\eta_B = \frac{J_Q}{2\pi} \frac{K_{TM}}{K_{QM}} \tag{5.26}$$

$$\eta_O = \frac{J_Q}{2\pi} \frac{K_T}{K_Q} \tag{5.27}$$

As can be seen from Fig. 5.21, by definition in this case: $K_{QM} = K_Q$ (torque identity), and $K_T = K_{TQ}$. Thus, the relative rotative efficiency obtained using the torque identity method is given by Eq. 5.28.

$$\eta_{RQ} = \frac{K_{TM}}{K_{TQ}} \tag{5.28}$$

Equation 5.28 is the ratio of the thrust coefficient measured in the self-propulsion test to the thrust coefficient in open water at the same advance coefficient.

5.5.1.2 Hydro-Acoustic Performance

As noted in Chap. 7, in addition to cavitation, hydro-acoustic noise can be narrowband, where the energy is focused at a number of discrete frequencies, or broadband, where the energy is spread across a wide frequency range.

Narrowband noise is due to the fluctuating loading on the propulsor blades, caused by the blades passing through the uneven wake field. Thus, in order to conduct a physical model test to investigate narrowband noise it is necessary to test the propulsor in situ behind a model of the submarine hull, complete with all its appendages. Since the wake from the hull is dependent on Reynolds number, as large a Reynolds number as possible should be used. If the tests are conducted in a quiet water tunnel with a very low background noise, such as the tunnel operated by the French DGA shown in Fig. 5.22, then the noise at model scale can be measured. Care is required to avoid self-noise on the hydrophones.

Broadband noise is due to turbulent interaction over the blade. Tests to determine broadband noise are generally conducted at as large a scale as possible, often in “open water”, without a submarine hull. Again, a facility like that shown in Fig. 5.22 can be used.

As noted in Chap. 7, the fluctuating forces on the propulsor can set up resonances in the submarine, which in turn can generate radiated hydro-acoustic noise. To predict this noise it is necessary to measure the fluctuating forces in all three dimensions, and use this with a numerical model of the submarine shaft and relevant structure. To do this a sophisticated dynamometer is required, capable of measuring at very high frequencies. This is because, due to the scaling laws, model

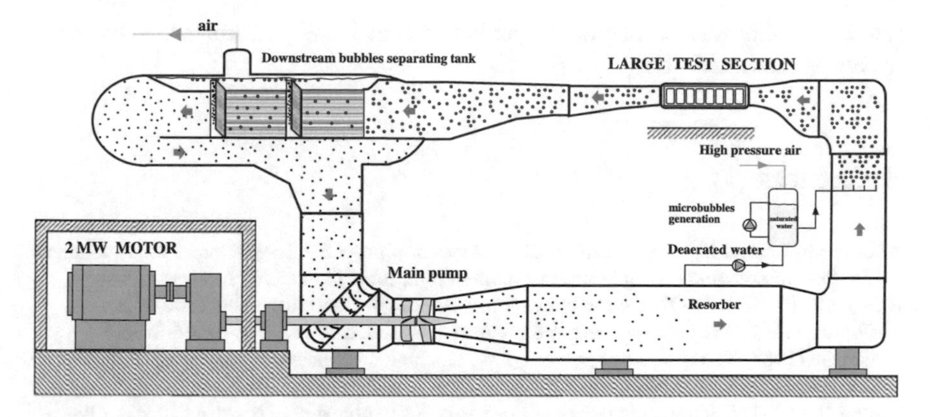

Fig. 5.22 Hydroacoustic and cavitation tunnel GTH (courtesy of DGA Hydrodynamics)

scale rpm is higher than full scale rpm, and it is necessary to capture the fluctuating forces at the higher harmonics. In addition, for a pumpjet the blade rate can be high, due to the number of stators and rotors.

5.5.2 Computational Fluid Dynamics

A range of Computational Fluid Dynamics (CFD) techniques exist which can be used for propulsor design. One of the simplest approaches is Reynolds-Averaged Navier-Stokes (RANS). The equations of motion of the flow are solved by splitting it into time-averaged and fluctuating components. This approach usually makes use of an empirically derived turbulence model. At present (2018) a number of commercial RANS solvers exist, and the use of RANS is considered to be a fairly routine approach to the simulation of fluid flow.

RANS techniques can be used very effectively to determine the wake flow into the propulsor. As noted above, this is important for the design of the propulsor.

RANS can also be used to predict the hydrodynamic performance of the propulsor itself, although at this time (2018) it is normal procedure, as a minimum, to conduct model experiments on the final design as confirmation.

As RANS can be used to predict both the wake field, and the propulsor hydrodynamic performance, in principle it can also be used to predict the fluctuating forces on the blades, and hence the narrowband hydro-acoustic performance. However, at this time (2018) this is generally not considered accurate enough to have confidence in this approach without model tests.

As RANS relies on a simplification of the boundary layer flow, it is not possible to use it to determine broadband hydro-acoustic noise. Further work is being undertaken using more sophisticated numerical approaches, however at this time

(2018) the only reliable method of predicting broadband performance is by model testing as discussed above.

References

Andersen P, Kappel JJ, Spangenberg E (2009) Aspects of propeller developments for a submarine. In: First International Symposium on Marine Propulsors, Trondheim, Norway, 2009

Burcher R, Rydill L (1998) Concepts in submarine design. Cambridge University Press

Carlton JS (2007) Marine propellers and propulsion. Elsivier. ISBN: 978-07506-8150-6

Clarke GE (1988) The choice of propulsor design for an underwater weapon. In: UDT conference, London, Oct 26–28, 1988

Dutton JL (1994) Contrarotating electric drive for attack submarines. Nav Eng J, Mar 1994

ITTC (2011a) Resistance tests, international towing tank conference recommended procedures and guidelines, Procedure number: 7.5-02-02-01

ITTC (2011b) Propulsion/bollard pull test, international towing tank conference recommended procedures and guidelines, Procedure number: 7.5-02-03-01.1

ITTC (2014) 1978 ITTC performance prediction method, international towing tank conference recommended procedures and guidelines, Procedure number: 7.5-02-03-01.4

Kormilitsin YN, Khalizev OA (2001) Theory of Submarine Design. Saint Petersburg State Maritime Technical University, Russia

McCormick BW, Eisenhuth J (1963) Propellors and pumpjets for underwater propulsion. AIAA J 1(10):2348–2354

Seil GJ, Anderson B (2012) A comparison of submarine fin geometry on the performance of a generic submarine. In: Proceedings of pacific 2012 international maritime conference, Sydney, 2012

SSPA (1993) Pumpjet propulsion. SSPA highlights magazine no. 2, 1993

van der Ploeg A (2012) Objective functions for optimizing a ship's aft body. In: Proceedings of the 11th international conference on computer and IT applications in the maritime industries (COMPIT), Liège, Belgium, pp 494–507, Apr 2012

van der Ploeg A (2015) RANS-based optimization of the aft part of ships including free surface effects. In: Proceedings of the international conference on computational methods in marine engineering, MARINE 2015, 15–17 June, Rome. pp 242–253

van Lammeren WPA, van Manen JD, Oosterveld MWC (1969) The Wageningen B screw series. Society of naval architects and marine engineers—transactions, vol 77

Vinton PM, Banks S, West M (2005) Astute propulsor technical innovation summary. In: Proceedings of warship 2005—naval submarines, Royal Institution of Naval Architects, London, 2005

Chapter 6
Appendage Design

Abstract Submarines usually have three groups of appendages: sail; forward control surfaces; and aft control surfaces. Appendages contribute a considerable increase in drag, and need to be considered carefully. There are two approaches to sail design: a foil type; and a blended type. The blended type of sail has a larger volume than the foil type, and is better faired into the hull. If the sail is at an angle of attack it will generate a side force high up resulting in a heel angle (particularly snap roll) and a force and moment in the vertical plane on the hull, resulting in a stern dipping tendency. The magnitude of the side force when manoeuvring will depend on the distance of the sail from the Pivot Point. The location of the sail will also affect the turning radius. The forward planes can be located in three different positions: midline; eyebrow; and sail. The pros and cons of each of these are discussed. The aft control surfaces may include fixed and movable surfaces, with the fixed surfaces increasing stability, and the movable surfaces used to change trim, and hence to make large depth changes, and to turn the submarine. Different aft control surface configurations include: cruciform; X-form; inverted Y; and pentaform. The pros and cons of these different configurations are discussed.

6.1 General

Submarines usually have three groups of appendages as follows:

(a) *sail*—to house periscopes, snorkel, and other masts, as well as to serve as the conning position when on the surface;
(b) *forward control surfaces*—to permit the submarine to change depth without changing trim, and to control depth at low speeds; and
(c) *aft control surfaces*—to control depth by changing trim, and to provide steering in the horizontal plane.

Appendages contribute a considerable increase in drag, and need to be considered carefully. They will often operate at an angle to the flow, thereby resulting in induced drag, and associated vortices. Care needs to be taken to ensure that root

M. Renilson, *Submarine Hydrodynamics*,
https://doi.org/10.1007/978-3-319-79057-2_6

fairing is done carefully, particularly over regions of the hull where the cross section is decreasing (see Sect. 4.1).

In addition, appendages need to be designed for the expected angle of the flow over them. This is of particular importance to the design of eyebrow planes, as noted in Sect. 6.3.3. However, it should also be a consideration for the aft control surfaces in an X-form configuration, as noted in Sect. 6.4.3.

The size of the control surfaces may be dictated by the need to control depth when at periscope depth in waves.

Control surfaces can be all moving, or can have a fixed part and a movable part. The purpose of the fixed part is generally to increase directional stability, thus this configuration is not common for forward control surfaces.

For control surfaces with a fixed and moving part the leading edge of the moving part can be directly connected to the trailing edge of the fixed part or there can be a gap between them, as shown in Fig 6.1. The former has the advantage of less turbulent flow in the straight ahead condition with zero control surface deflection. However, as the movable part is hinged at its leading edge this requires a large torque. If a gap is provided between the movable part and the fixed part there can be flow between them when the submarine is operating at large angles of attack, which may be beneficial. Also, the movable part can be hinged some way from its leading edge, resulting in a lower required torque.

In addition to control surfaces it is sometimes necessary to add further fixed appendages aft to improve a submarine's directional stability. For an X-form configuration this can be used to improve stability in the vertical plane, as shown in Fig. 6.2.

Predicting the drag on appendages using physical model experiments is particularly difficult due to their small chords, and hence very low Reynolds numbers at model scale.

Fig. 6.1 Aft control surface with gap between fixed and moving parts

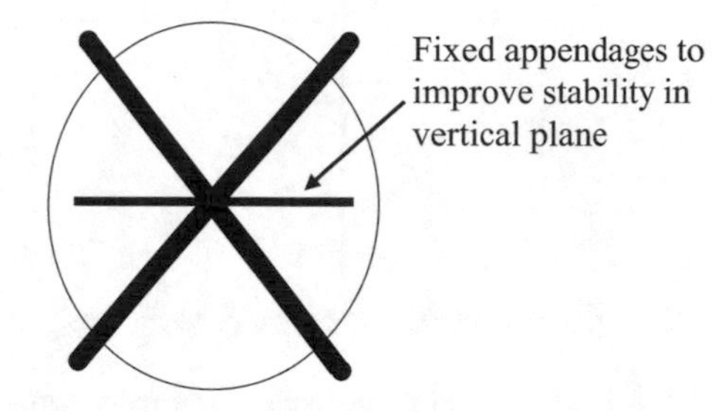

Fig. 6.2 Fixed appendage to improve stability in the vertical plane

6.2 Sail

The sail, or bridge fin, is undesirable from a hydrodynamic point of view, as it adds to the drag on the submarine, adversely affects the flow into the propulsor, and has a detrimental effect on manoeuvring in the horizontal plane, as discussed in Sect. 3.6.

The further forward the sail is placed the smaller the hull boundary layer that it will be operating in, and hence the lower the interference drag (Hoerner 1965). In addition, the further forward the sail is the better the flow is likely to be into the propulsor. However, the position of the sail also influences the manoeuvring characteristics, as discussed in Sect. 3.6.

There are two types of sail, the foil type, and the blended type (or sedan) as shown in Fig. 6.3 adapted from Seil and Anderson (2012), provided by the authors.

The strategy with the foil type is to reduce the size of the sail as much as possible. Thus, large US nuclear submarines have relatively small sails compared to the size of their hulls. However, the recent trend to use the sail to accommodate equipment to be used by special forces, together with the need for a larger number of masts etc, has meant that the size of the sail on some modern submarines is now larger than previously.

It is important to ensure that where the sail meets the hull it is faired as well as possible to reduce the magnitude of vortices, which can adversely affect the flow into the propulsor, and influence the manoeuvring in the horizontal plane

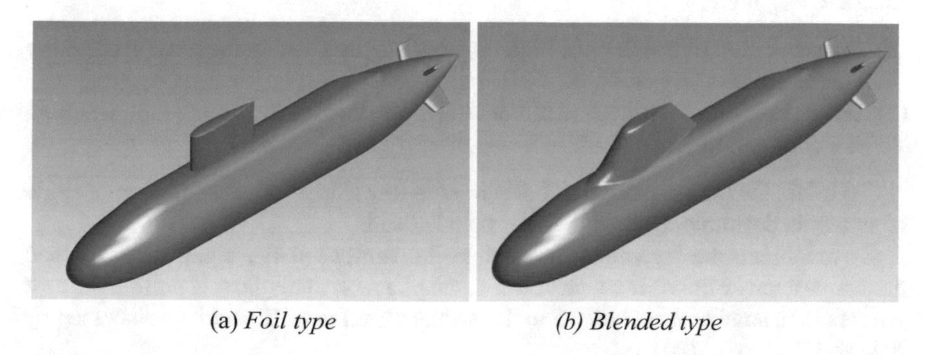

(a) *Foil type* (b) *Blended type*

Fig. 6.3 Types of sail (adapted from Seil and Anderson 2012)

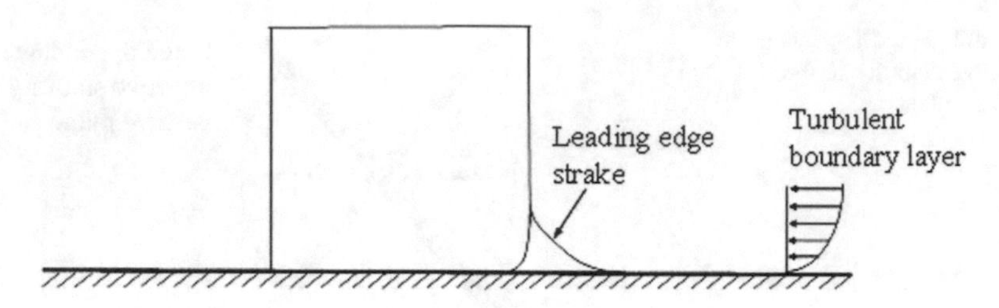

Fig. 6.4 Leading edge strake on sail to reduce horseshoe vortices

(Sect. 3.6). A leading edge strake, as shown in Fig. 6.4, can be provided to reduce the magnitude of the horseshoe vortices. This is discussed in more detail in Sect. 4.7.

Care also needs to be taken with the shape of the tip of the sail, to ensure that the magnitude of tip-vortex is reduced. The increased drag due to interaction with the hull can be reduced if the sail is at a distance of 0.2–0.3 of the hull length from the forward perpendicular (Kormilitsin, and Khalizev 2001).

The blended type of sail has a larger volume than the foil type, and is better faired into the hull, reducing the effect of root vortices. However, the greater volume may result in increased drag, including wave drag when operating near the surface. Transverse stability when surfacing may also be affected, depending on the drainage arrangements for the free flooding compartments in the sail.

Seil and Anderson (2012) pointed out that a blended sail can be designed to reduce the overall drag, as although the drag on the sail is increased, the total wetted surface of the hull + sail can be reduced, as illustrated in Fig. 6.5. This is for a constant sail height.

They also showed that although the wake into the propulsor from a badly designed blended sail can be considerably worse than that from a foil sail, the wake into the propulsor from a well-designed blended sail can be equivalent to that from a foil sail.

If the sail is at an angle of attack it will adversely affect the manoeuvring in the horizontal plane, as discussed in Sect. 3.6. The two primary adverse effects on manoeuvring are:

(a) the generation of a side force high up, resulting in a heel in the turn (particularly snap roll); and
(b) the generation of a force and moment in the vertical plane on the hull, resulting a stern dipping tendency.

To minimize the snap roll a sail design with low side force as a function of angle of attack is desirable, such as a small blended sail.

To minimize the force and moment in the vertical plane, a sail design which generates a lower tip-vortex when at an angle of attack to the flow is preferable, as it will result in smaller modification to the vortices shed from the hull, as discussed by Seil and Anderson (2013).

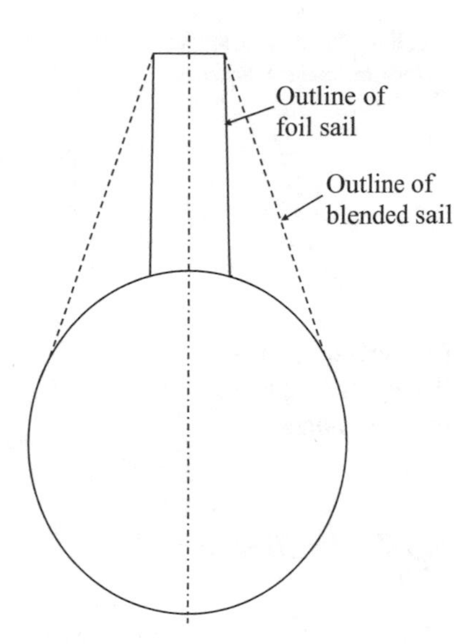

Fig. 6.5 Comparison of fin cross sections

Thus, the design of the sail depends on many factors, which need to be considered at the early design stage.

As noted in Sect. 3.6 locating the sail at the Pivot Point will reduce the local angle of attack on it when the submarine is manoeuvring in the horizontal plane, thereby reducing the side force. In the past the longitudinal location of the sail was required to be above the control room, as it was necessary for the periscopes to be located there, however with modern non-penetrating masts this restriction no longer applies.

Sails which are placed well forward will reduce the turning radius, whereas sails placed aft will increase it.

6.3 Forward Control Surfaces

6.3.1 General

Forward control surfaces are required to enable the submarine to change depth without changing trim, which is important at periscope depth. They are also necessary to provide control in the vertical plane at low speeds, as discussed in Sect. 3.7.

See Sect. 3.9 for recommended values of forward plane effectiveness.

There are three possible locations for the forward control surfaces, as shown in Fig. 6.6:

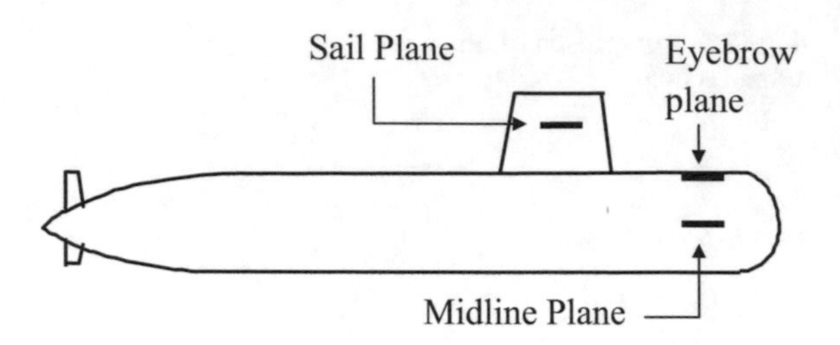

Fig. 6.6 Possible locations of forward control surfaces

(a) midline planes;
(b) eyebrow planes; and
(c) sail planes.

6.3.2 Midline Planes

Midline planes are in reasonably undisturbed flow, and the hull provides a positive ground-board effect, thus increasing their effective aspect ratio, as shown in Fig. 6.7.

However, with midline planes the trailing-vortices may degrade the performance of the sonar arrays along the side of the hull (flank arrays) as shown in Fig. 6.8. In addition, the vortices may be sucked into the propulsor, increasing propulsor noise, also shown in Fig. 6.8.

Midline planes need to be retractable. This also gives the opportunity to reduce resistance and noise associated with them, as the forward planes are not needed when operating deep and at high speed, as discussed in Sect. 3.7.

6.3.3 Eyebrow Planes

Eyebrow planes operate in the upward flow caused by the hull, as shown in Fig. 6.9. The upward flow will be stronger closer to the hull. This means that unless

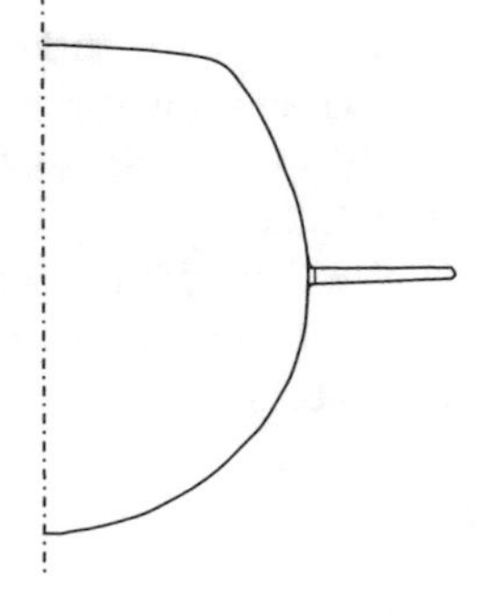

Fig. 6.7 Cross section of midline plane

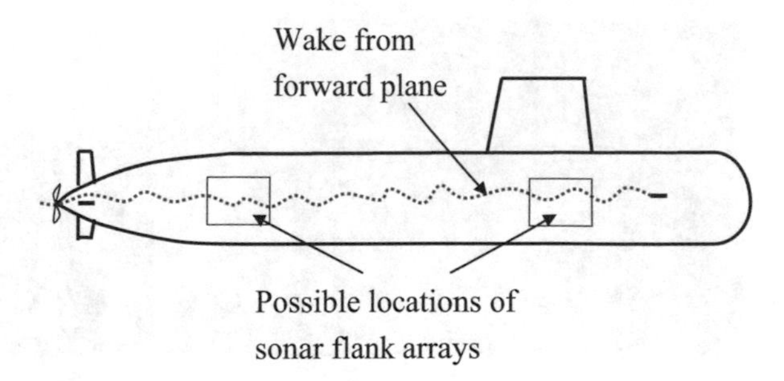

Fig. 6.8 Wake from midline plane

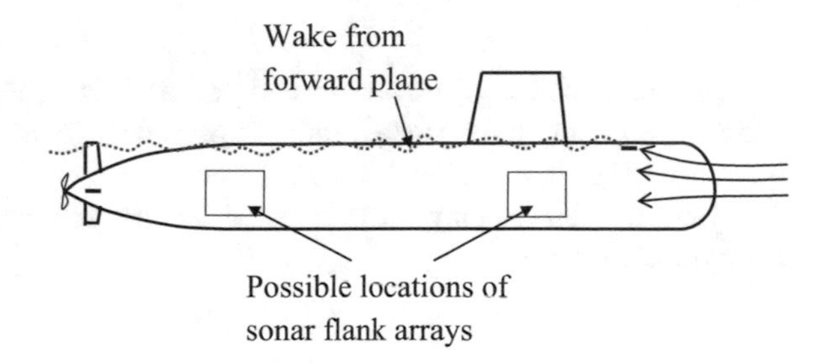

Fig. 6.9 Eyebrow plane, showing upward flow due to the hull, and the wake from the plane

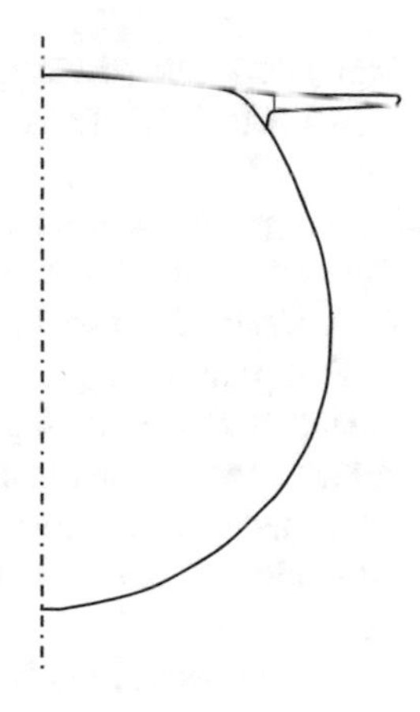

Fig. 6.10 Cross section of eyebrow plane

the plane is twisted in the span-wise direction the angle of attack on the plane will be a function of the span-wise position.

Thus, it will not be possible for the whole plane to be at zero angle of attack, and hence there will always be induced drag and associated tip-vortices. In addition, great care will be required with the root fillet and interaction with the hull as shown in Fig. 6.10. In many cases, there will be a gap when the plane is at a non-zero angle of attack, and this can result in vortices and associated noise. As seen in Fig. 6.11, one solution is to modify the hull in way of the eyebrow plane, such that the hull is flat in the region of the connection between the plane root and the hull.

Fig. 6.11 Photographs of eyebrow plane on submarine model

With the eyebrow plane the tip-vortices are less likely to affect the flank arrays, and are less likely to be sucked into the propulsor than with midline planes, as can be seen in Fig. 6.9.

Eyebrow planes may or not need to be retractable, depending on their configuration.

6.3.4 Sail Planes

Sail planes are situated in reasonably undisturbed flow. The vortices from them won't affect the flank arrays and are unlikely to be sucked into the propulsor, Fig. 6.12.

Sail planes will have a good ground-board effect, thus increasing their effective aspect ratio, as shown in Fig. 6.13. Sail planes do not need to be retractable, as they are contained completely inside the overall dimensions of the submarine.

Sail planes are generally located close to the Neutral Point (see Sect. 3.7). This means that by operating the sail planes alone it is possible to change the depth of the submarine without changing its trim. With earlier, manual, control systems this may have been useful, but with modern control techniques it is easy to apply a net vertical force at the Neutral Point from two well-spaced planes, making this feature less important. However, not having to operate the aft control surfaces when at periscope depth may be an advantage.

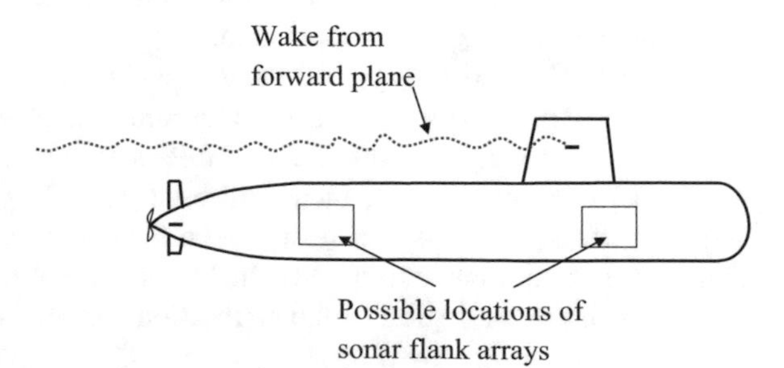

Fig. 6.12 Wake from sail plane

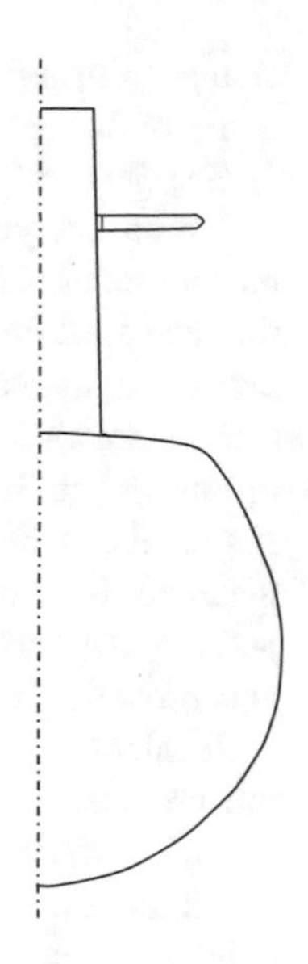

Fig. 6.13 Cross section of sail plane

On the other hand, as they are not as far forward as either the midline, or the eyebrow planes, they have less lever arm to the Critical Point, making them less effective when depth changes with trim are required.

As sail planes are located much higher than the hull, and relatively close to the surface when operating at periscope depth, they will be influenced by the presence of wind generated waves. This may need to be taken into account when assessing a submarine's performance at periscope depth in waves.

In addition, the mass of the sail plane, and associated machinery, which are situated high up on the boat may cause stability issues, particularly when passing through the surface, as discussed in Chap. 2.

Sail planes are less effective in assisting a submarine to crash dive than those placed on the hull. However, with modern submarines which normally operate submerged, and are not required to crash dive, this is usually not important. Sail places pose complications for submarines which are required to be able to surface through ice. It is possible to do this if plane can rotate to the vertical, however it is simpler to avoid the use of sail planes for such boats if possible.

6.4 Aft Control Surfaces

6.4.1 General

The aft control surfaces may include fixed and movable surfaces. The movable surfaces, stern planes and rudders, are required to change trim, and hence to make large depth changes, and to turn the submarine. They are also used to control depth changes during a turn. The fixed surfaces are provided to increase stability, particularly in the vertical plane, if required.

See Sect. 3.9 for recommended values of stern plane and rudder effectiveness.

On an axisymmetric body with a single propulsor the aft control surfaces are usually located forward of the propulsor. This means that they do not benefit from the increased flow from the propulsor, and that they cause wake distortions into the propulsor, increasing propulsor noise. Thus the number of aft control surfaces, and their interaction with the wake from the sail, needs to be considered along with the propulsor blade number when considering their effect on acoustic signature. A propeller with the same number of blades as the number of wake regimes would generate significant acoustic noise, as each of the blades would be passing through a wake at the same time. This should be avoided, as should any multiples of wake and propeller blade numbers.

Ideally the aft control surfaces should be moved as far forward, and away from the propulsor, as possible. However, this will reduce the available span for the control surfaces, due to the increase in tail cone diameter, and the preference to avoid them protruding beyond the dimensions of the submarine, to avoid complications when coming alongside, as shown in Fig. 6.14. Note that for many submarines the need to provide adequate stability in the vertical plane results in the stern planes extending beyond the side of the hull—i.e. outside the desirable "bucket" shown in Fig 6.14.

It is important to note that as the submarine is symmetrical in the *x-z* plane, manoeuvres in the vertical plane only require a vertical force from the aft control surfaces. On the other hand, as it is not symmetrical in the *x-y* plane, manoeuvres in the horizontal plane will require both a force in the horizontal direction and one in the vertical direction to maintain constant depth (see Sect. 3.6). Thus, when considering the size and configuration of the aft control surfaces this needs to be taken into account. There is little point in providing a large horizontal force capability if the vertical force available cannot maintain the depth when turning at maximum turn rate.

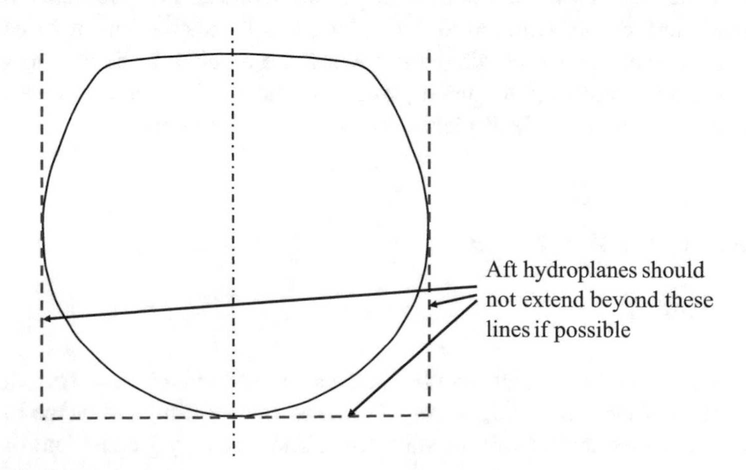

Fig. 6.14 U shape "bucket" defining ideal limits of planes

6.4.2 Cruciform Configuration

The traditional arrangement for the aft control surfaces on submarines with a single axial propulsor is the cruciform configuration, as shown in Figs. 6.15 and 6.16. With this arrangement the vertical control surfaces (rudders) control the manoeuvring in the horizontal plane, and the horizontal control surfaces (stern planes) control the manoeuvring in the vertical plane. As the submarine is symmetrical in the x-z plane, manoeuvres in the vertical plane only require operation of the stern planes, whereas as it is not symmetrical in the x-y plane, manoeuvres in the horizontal plane will require operation of the rudders, together with operation of the stern planes (and bow planes) to maintain constant depth (see Sect. 3.6).

As discussed in Sect. 3.1 it is normally desirable for a submarine to have a high degree of stability in the vertical plane, and a high degree of manoeuvrability (lower stability) in the horizontal plane. The cruciform arrangement makes it possible to achieve this. For example, it is possible to have a fixed fin with a flap for the stern planes, and an all moving arrangement for the rudders, as shown in Fig. 6.16a. Note that this is not essential, and many submarines have a fixed fin and flap for the rudder, as well as for the stern plane (see Appendix).

With the cruciform configuration the lower rudder is often smaller, with a lower aspect ratio, than desirable as it is not ideal for it to extend below the keel of the submarine. This may make manoeuvring on the surface difficult. The upper rudder may be made larger, which will also help reduce the snap roll when submerged, but as the upper rudder operates in the wake from the sail it will be less effective.

Fig. 6.15 Cruciform configuration for aft planes (courtesy QinetiQ Limited, © Copyright QinetiQ Limited 2017)

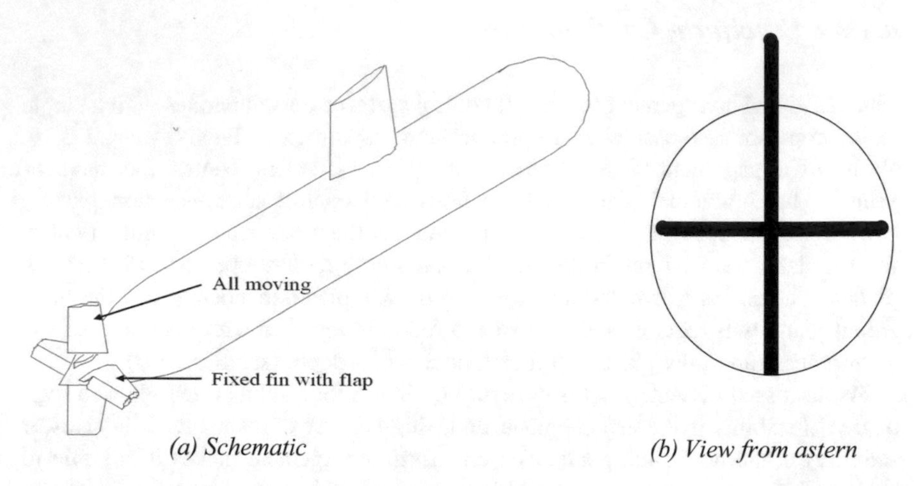

Fig. 6.16 Cruciform configuration for aft planes

The span of the stern planes is also limited by the desire for them not to protrude beyond the beam of the submarine as shown in Fig. 6.14. However, to achieve the required stability in the vertical plane, G_V, and the stern plane effectiveness, given in Table 3.9, this is not always possible. Fig. 6.17 is a sketch of the stern planes sized to meet the requirements in Table 3.9. As can be seen, if the stern planes are permitted to extend beyond the hull boundary they can have a reasonably high aspect ratio as can be seen in Fig. 6.17a. However, if required to remain within the hull boundary they are likely to have a very low aspect ratio, as shown in Fig. 6.17b.

The stern planes in a cruciform configuration can also be fitted with vertical end plates as shown in Fig. 6.18. These have the effect of increasing the vertical area aft, therefore increasing stability in the horizontal plane, and increasing the effectiveness of the stern planes, due to the increased effective aspect ratio as a result of the addition of the end plate. However, these additional end plates increase the wetted surface, and hence the drag (see Chap. 4). They also increase the mass and complexity of the stern configuration.

A major disadvantage of the cruciform configuration exists when the two stern planes are connected to each other, and the two rudders are connected to each other. This is often done to reduce mechanical complexity and mass in the stern. Thus, if there is a jam in the stern planes, for example, this cannot be recovered by moving the other stern plane. See Sect. 3.10 for a discussion of Safe Operating Envelopes. If the control surfaces can be moved independently of each other, then this limitation would be reduced, however the complexity, and mass, of the linkages within the stern required to do this are significantly increased.

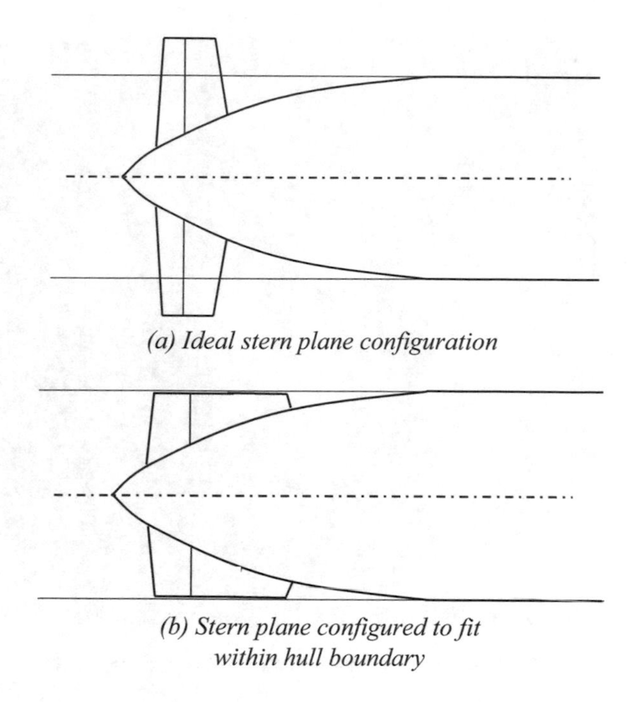

Fig. 6.17 Plan view of alternative stern plane shapes for cruciform configuration

6.4.3 X-form Configuration

An alternative arrangement for the aft control surfaces is the X-form configuration, as shown in Figs. 6.19 and 6.20. With this configuration, as each control surface is identical, it is not so easy to meet the generally accepted requirement of a high level of stability in the vertical plane together with a lower level of stability in the horizontal plane to permit good maneuverability. The need for a high level of stability in the vertical plane may be less important for low speed conventional submarines (SSKs) than for the higher speed nuclear powered submarines (SSNs). Control strategies may be used to overcome the hydrodynamic limitations, however this aspect does need to be considered with the X-form configuration.

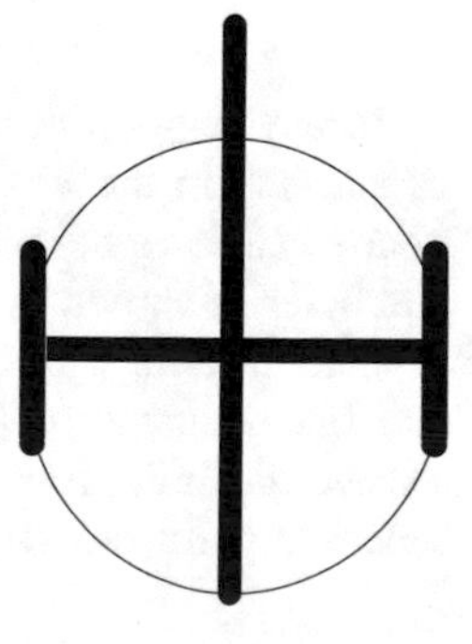

Fig. 6.18 Cruciform configuration with end plates on stern planes

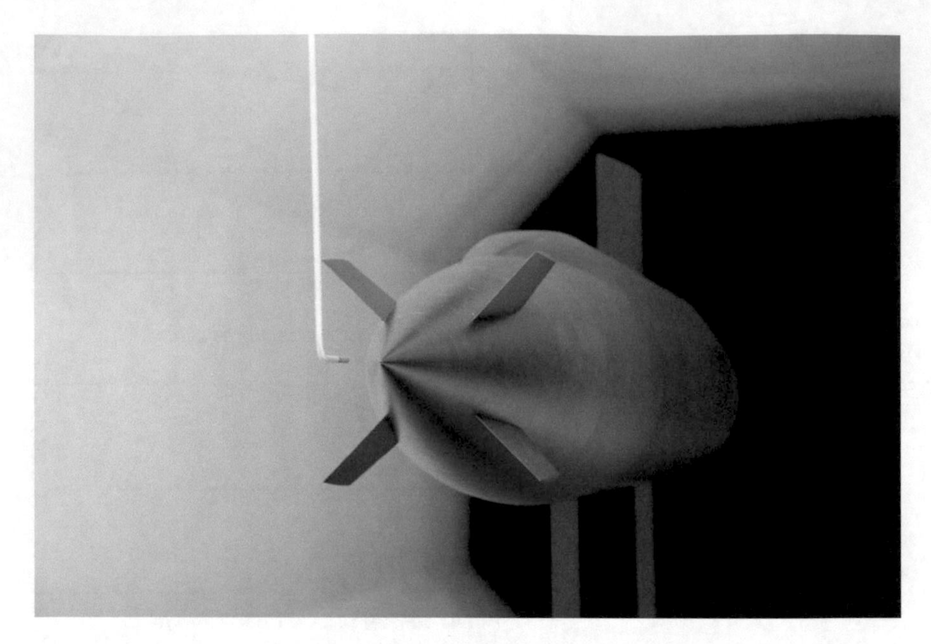

Fig. 6.19 X-form configuration for aft planes (courtesy DST Group)

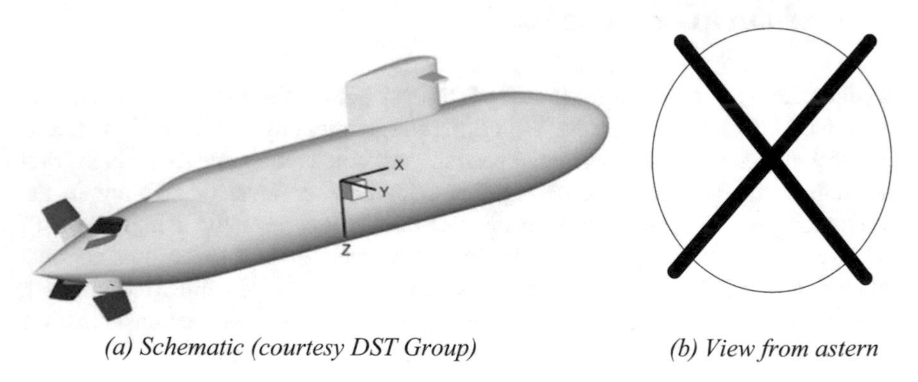

(a) Schematic (courtesy DST Group) *(b) View from astern*

Fig. 6.20 X-form configuration for aft planes

It may be possible to go some way towards the requirement for different degrees of stability in the vertical and horizontal planes by not positioning the individual control surfaces at an angle of 90° to each other, and/or adding a horizontal fixed stabilizer as shown in Fig. 6.2.

One advantage of the X-form configuration is that each of the control surfaces can have a much longer span than with the cruciform configuration before they exceed the limit given by the "bucket" defined by the dashed line in Fig. 6.14. This means that for the same aspect ratio the total control surface area for the X-form

Fig. 6.21 Control inputs for X-form configuration (taken from Renilson 2011)

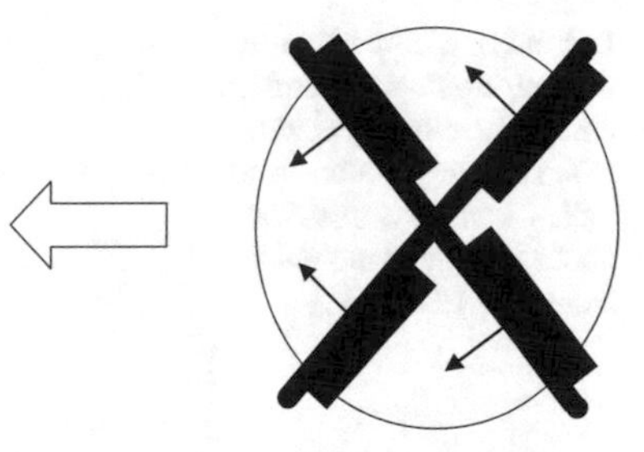

(a) Rudder for turn to starboard

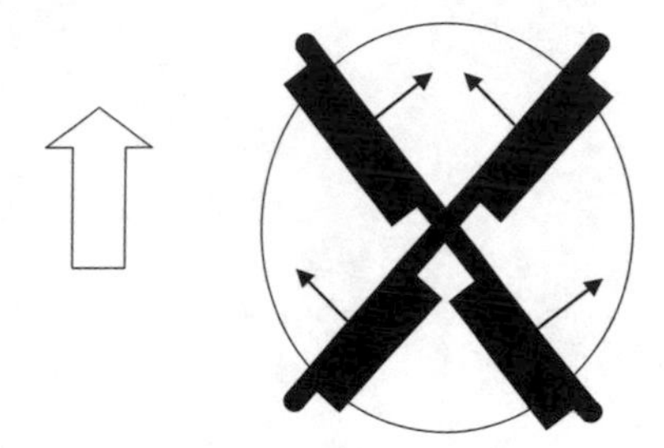

(b) Stern planefor stern to rise

configuration can be much larger than that for a cruciform configuration. Or, alternatively, the aspect ratio can be increased for the same total control surface area. Either way, there is the opportunity of generating greater control forces with the X-form configuration, than with the cruciform configuration. This can result in substantially better manoeuvring on the surface for the X-form configuration than with the cruciform configuration, particularly if the lower rudder in the cruciform configuration is not permitted to extend below the keel line.

With the X-form configuration, generally each of the control surfaces is all moving and completely independent. This requires an autopilot to control the submarine, as for each manoeuvre a different combination of plane movements is required. As can be seen from Fig. 6.21a, when applying a force to port, in order to turn to starboard, each control surface is required to operate. Vertical upwards forces are generated which are cancelled out by vertical downwards forces. Hence, to create a given *effective* force considerably greater *total* force is generated for the X-form configuration than for the cruciform configuration. The same can be seen in Fig. 6.21b when creating a vertical force (Renilson 2011).

Thus, a greater total force will be required for the X-form configuration. This will result in greater formation of vortices which will impinge on the propulsor, increasing noise. It will also result in greater induced drag, and hence more power will be required to maintain a given speed than for the equivalent cruciform configuration.

However, as most submarines are not symmetrical about the x-y plane, when turning a vertical force is often required to maintain depth (Sect. 3.6). With the

Fig. 6.22 Comparison of control inputs for a cruciform and X-form configurations when turning a submarine which is not symmetrical about the *x-y* plane (taken from Renilson 2011)

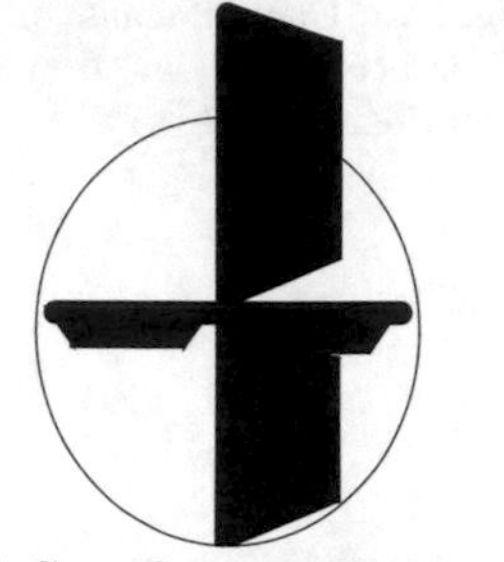

(a) Cruciform configuration

(b) X configuration

cruciform arrangement this means that for a level turn the stern planes are used in addition to the rudder. For the X-form configuration each individual plane already provides a vertical force, so, depending on the level of asymmetry, the *total* force may not be greater for the X-form configuration, as illustrated in Fig. 6.22. Whether the *total* force is greater for the cruciform or the X-form for a particular manoeuvre will depend on the degree of asymmetry that the submarine has.

As a submarine requires a vertical force when turning to compensate for the asymmetry, the magnitude of the vertical force available from the aft control surfaces may be the effective limit to the rate of turn, rather than the magnitude of the horizontal force. For example, with the X-form configuration adopted by Crossland et al. (2011, 2012) the maximum effective rudder angle is about 25°, with greater angles not being possible due to the limit in the vertical force available.

For an X-form configuration the forces and moments generated by each individual planes need to be considered in the representations of the hydrodynamic forces and moments, Eqs. 3.7–3.12. A new set of terms for the aft planes is required, as given in Table 6.1, adapted from Crossland et al. (2011, 2012).

In Table 6.1b the subscript X_i refers to X-form plane number "i".

Following the same approach as presented in Eqs. 3.7–3.12, neglecting interaction between the planes and using subscripts 1–4 to designate each of the individual control surfaces Eqs. 6.1–6.6 are obtained.

Table 6.1 Comparison of cruciform and X-form appendage coefficients (taken from Crossland et al. 2011)

(a) Cruciform configuration	
Rudder	Stern plane
$X'_{\delta R \delta R}$	$X'_{\delta S \delta S}$
$Y'_{\delta R}$	–
–	$Z'_{\delta S}$
$K'_{\delta R}$	–
–	$M'_{\delta S}$
$N'_{\delta R}$	–

(b) X-form configuration
$X'_{\delta X \delta X_i}$
$Y'_{\delta X_i}$
$Z'_{\delta X_i}$
$K'_{\delta X_i}$
$M'_{\delta X_i}$
$N'_{\delta X_i}$

$$\begin{aligned} X = \frac{1}{2}\rho L^2 \big[& X'_{\delta 1 \delta 1} u^2 \delta_1^2 + X'_{\delta 2 \delta 2} u^2 \delta_2^2 + X'_{\delta 3 \delta 3} u^2 \delta_3^2 + X'_{\delta 4 \delta 4} u^2 \delta_4^2 \\ & + \left(X'_{\delta 1 \delta 1 \eta} u^2 \delta_1^2 + X'_{\delta 2 \delta 2 \eta} u^2 \delta_2^2 + X'_{\delta 3 \delta 3 \eta} u^2 \delta_3^2 + X'_{\delta 4 \delta 4 \eta} u^2 \delta_4^2 \right)(\eta - 1) \big] \end{aligned} \tag{6.1}$$

$$\begin{aligned} Y = \frac{1}{2}\rho L^3 & \Big[Y'_{|r|\delta 1} u|r|\delta_1 + Y'_{|r|\delta 2} u|r|\delta_2 + Y'_{|r|\delta 3} u|r|\delta_3 + Y'_{|r|\delta 4} u|r|\delta_4 \\ & + Y'_{|q|\delta 1} u|q|\delta_1 + Y'_{|q|\delta 2} u|q|\delta_2 + Y'_{|q|\delta 3} u|q|\delta_3 + Y'_{|q|\delta 4} u|q|\delta_4 \Big] \\ & + \frac{1}{2}\rho L^2 \big[Y'_{\delta 1} u^2 \delta_1 + Y'_{\delta 2} u^2 \delta_2 + Y'_{\delta 3} u^2 \delta_3 + Y'_{\delta 4} u^2 \delta_4 \\ & + (Y'_{\delta 1 \eta} u^2 \delta_1 + Y'_{\delta 2 \eta} u^2 \delta_2 + Y'_{\delta 3 \eta} u^2 \delta_3 + Y'_{\delta 4 \eta} u^2 \delta_4)(\eta - 1) \Big] \end{aligned} \tag{6.2}$$

$$\begin{aligned} Z = \frac{1}{2}\rho L^3 & \Big[Z'_{|r|\delta 1} u|r|\delta_1 + Z'_{|r|\delta 2} u|r|\delta_2 + Z'_{|r|\delta 3} u|r|\delta_3 + Z'_{|r|\delta 4} u|r|\delta_4 \\ & + Z'_{|q|\delta 1} u|q|\delta_1 + Z'_{|q|\delta 2} u|q|\delta_2 + Z'_{|q|\delta 3} u|q|\delta_3 + Z'_{|q|\delta 4} u|q|\delta_4 \Big] \\ & + \frac{1}{2}\rho L^2 \big[Z'_{\delta 1} u^2 \delta_1 + Z'_{\delta 2} u^2 \delta_2 + Z'_{\delta 3} u^2 \delta_3 + Z'_{\delta 4} u^2 \delta_4 \\ & + \left(Z'_{\delta 1 \eta} u^2 \delta_1 + Z'_{\delta 2 \eta} u^2 \delta_2 + Z'_{\delta 3 \eta} u^2 \delta_3 + Z'_{\delta 4 \eta} u^2 \delta_4 \right)(\eta - 1) \Big] \end{aligned} \tag{6.3}$$

$$\begin{aligned} K = \frac{1}{2}\rho L^3 \big[& K'_{\delta 1} u^2 \delta_1 + K'_{\delta 2} u^2 \delta_2 + K'_{\delta 3} u^2 \delta_3 + K'_{\delta 4} u^2 \delta_4 \\ & + \left(K'_{\delta 1 \eta} u^2 \delta_1 + K'_{\delta 2 \eta} u^2 \delta_2 + K'_{\delta 3} \eta u^2 \delta_3 + K'_{\delta 4} \eta u^2 \delta_4 \right)(\eta - 1) \Big] \end{aligned} \tag{6.4}$$

$$
\begin{aligned}
M = &\frac{1}{2}\rho L^4 \Big[M'_{|r|\delta 1}u|r|\delta_1 + M'_{|r|\delta 2}u|r|\delta_2 + M'_{|r|\delta 3}u|r|\delta_3 + M'_{|r|\delta 4}u|r|\delta_4 \\
&+ M'_{|q|\delta 1}u|q|\delta_1 + M'_{|q|\delta 2}u|q|\delta_2 + M'_{|q|\delta 3}u|q|\delta_3 + M'_{|q|\delta 4}u|q|\delta_4\Big] \\
&+ \frac{1}{2}\rho L^3 \big[M'_{\delta 1}u^2\delta_1 + M'_{\delta 2}u^2\delta_2 + M'_{\delta 3}u^2\delta_3 + M'_{\delta 4}u^2\delta_4 \\
&+ \Big(M'_{\delta 1\eta}u^2\delta_1 + M'_{\delta 2\eta}u^2\delta_2 + M'_{\delta 3\eta}u^2\delta_3 + M'_{\delta 4\eta}u^2\delta_4\Big)(\eta - 1)\Big]
\end{aligned} \tag{6.5}
$$

$$
\begin{aligned}
N = &\frac{1}{2}\rho L^4 \Big[N'_{|r|\delta 1}u|r|\delta_1 + N'_{|r|\delta 2}u|r|\delta_2 + N'_{|r|\delta 3}u|r|\delta_3 + N'_{|r|\delta 4}u|r|\delta_4 \\
&+ N'_{|q|\delta 1}u|q|\delta_1 + N'_{|q|\delta 2}u|q|\delta_2 + N'_{|q|\delta 3}u|q|\delta_3 + N'_{|q|\delta 4}u|q|\delta_4\Big] \\
&+ \frac{1}{2}\rho L^3 \big[N'_{\delta 1}u^2\delta_1 + N'_{\delta 2}u^2\delta_2 + N'_{\delta 3}u^2\delta_3 + N'_{\delta 4}u^2\delta_4 \\
&+ \Big(N'_{\delta 1\eta}u^2\delta_1 + N'_{\delta 2\eta}u^2\delta_2 + N'_{\delta 3\eta}u^2\delta_3 + N'_{\delta 4\eta}u^2\delta_4\Big)(\eta - 1)\Big]
\end{aligned} \tag{6.6}
$$

Now it is necessary to be able to obtain the values of the coefficients for each of the individual aft control surfaces. It is not possible to assume that these are the same for each coefficient. It is necessary to obtain the forces and moments as a result of individual plane deflections to populate Eqs. 6.1–6.6, however, it is also desirable to conduct a limited investigation into potential interaction between them, particularly when the submarine is operating at an angle of attack to the flow, Pook et al. (2017).

It should be noted that the lower pair of planes can be slightly more effective than the upper pair, possibly due to the presence of the sail creating vortices and/or the presence of the casing which impact the upper planes. Also, the interference between planes is negligible at low plane angles, but may become noticeable at plane angles above about 15° (Crossland et al. 2012).

Another issue with using the X-form configuration is that the flow into the planes will be affected by the asymmetrical shape of the aft body, including the sail and the casing. This means that if they are not aligned carefully they will experience a small angle of attack when the plane angle is zero (Pook et al. 2017). Although this can also be an issue with the stern planes on a cruciform configuration, it is less of an issue for that configuration. Of course, due to the symmetry, this is not an issue for the rudders in a cruciform configuration.

In an X-form configuration single plane jams at moderate angles should be able to be comfortably dealt with by using the other three planes. However, plane jams at higher angles may also require the submarine to slow down in order to control both heading and pitch changes (Crossland et al. 2012). This will require a different approach to the Safe Operating Envelopes, as discussed in Sect. 3.10.

6.4.4 Alternative Configurations

One possible alternative configuration is the inverted Y-form, as shown in Fig. 6.23.

With this configuration, when submerged manoeuvring in the horizontal plane is achieved by the single rudder alone, and the two lower planes are used for manoeuvring in the vertical plane. However, on the surface, the lower planes are required for manoeuvring in the horizontal plane.

There is no restriction on the rudder span, and the position of the rudder will reduce the snap roll, however its effectiveness may be influenced by the wake from the sail.

The requirement to have good stability in the vertical plane and good manoeuvrability in the horizontal plane can be achieved by sizing the rudder and the lower planes accordingly. It is also possible for the lower planes to have a fixed fin and a flap, rather than to be all moving. This will increase stability in the vertical plane, as with the stern planes in the cruciform configuration shown in Figs. 6.15 and 6.16.

With the inverted Y-form configuration the angles to the horizontal of the lower planes can be set to best suit the balance of manoeuvring and stability in the vertical and horizontal planes accordingly.

The inverted Y-form configuration results in three different wakes into the propulsor, which may affect the propulsor noise. The problems with the difficulty of aligning the upper control surfaces with the flow, associated with the X-form configuration as discussed above, do not arise, as the flow into the upper rudder will be purely axial, due to the asymmetry.

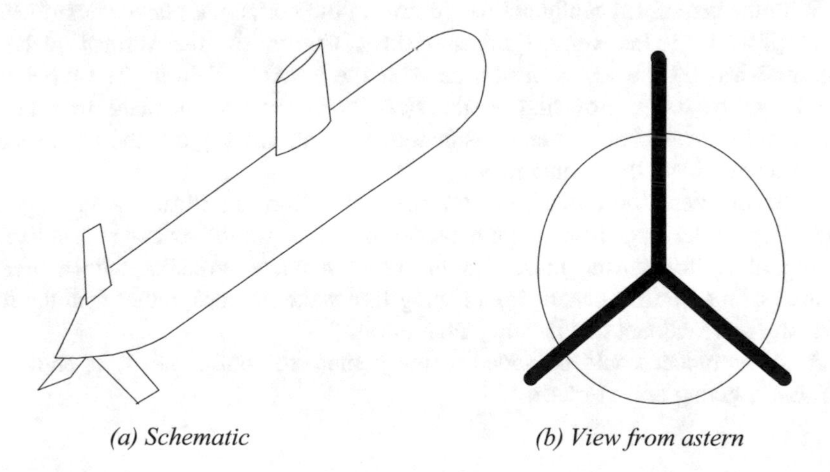

Fig. 6.23 Inverted Y-form configuration

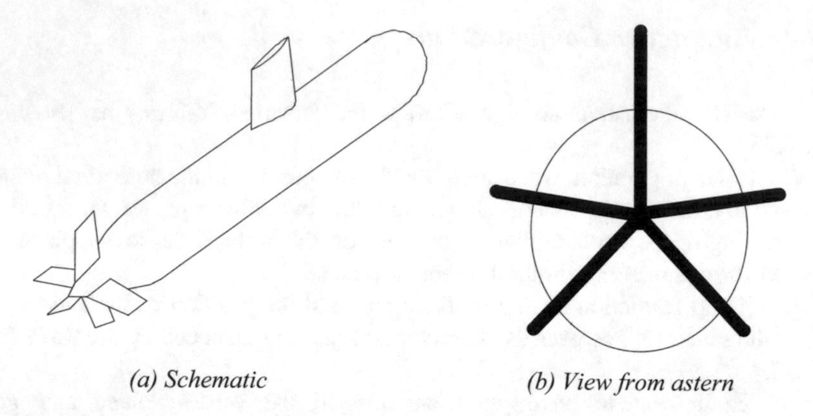

Fig. 6.24 Pentaform configuration

Fig. 6.25 Modified pentaform configuration with additional lower rudder

A further alternative, is the pentaform configuration, as shown in Fig. 6.24.

With the pentaform configuration, control in the horizontal plane when dived is accomplished by the vertical upper rudder. Control in the vertical plane is accomplished by the lower planes, as with the inverted Y-form. The other two planes can be fixed, providing greater stability in the vertical plane than in the horizontal plane, which is often a design requirement. The angle of these planes can be set to best meet this requirement.

With the pentaform configuration there are five significant wake regimes impacting on the propulsor, so a five bladed propeller would generate considerable additional hydro-acoustic noise. On the other hand, a propeller with a greater number of blades may benefit from having five wake regimes, rather than the four from the conventional cruciform configuration.

A lower rudder could be added to the pentaform configuration, as shown in Fig. 6.25, giving six aft planes.

References

Crossland P, Kimber NI, Thompson N (2012) Understanding the manoeuvring performance of an X-plane submarine in deep water and near the free surface. In: Pacific 2012: international maritime conference, Sydney, Jan 2012

Crossland P, Marchant P, Thompson N (2011) Evaluating the manoeuvring performance of an X-plane submarine. In: Proceedings of warship 2011: naval submarines and UUVs, Royal Institution of Naval Architects, Bath, 29–30 June 2011

Hoerner SF (1965) Fluid-dynamic drag

Kormilitsin YN, Khalizev OA (2001) Theory of submarine design. Saint Petersburg State Maritime Technical University, Russia

Pook DA, Seil G, Nguyen M, Ranmuthugala, D, Renilson MR (2017) The effect of aft control surface deflection at angles of drift and angles of attack. In: Proceedings of warship 2017: naval submarines and UUVs, Royal Institution of Naval Architects, Bath, UK

Renilson MR (2011) Submarine manoeuvring and appendage design—what is the best option for a large SSK? In: Proceedings of the Submarine Institute of Australia Technology conference 2011, Science Technology and Engineering, Adelaide, 8–10 Nov 2011

Seil GJ, Anderson B (2012) A comparison of submarine fin geometry on the performance of a generic submarine. In: Proceedings of Pacific 2012 international maritime conference, Sydney, 2012

Seil GJ, Anderson B (2013) The influence of submarine fin design on heave force and pitching moment in steady drift. In: Pacific 2013: international maritime conference, Sydney, Oct 2013

Chapter 7
Hydro-acoustic Performance

Abstract The propulsor is the most important source of hydrodynamic noise. Low rotational speed and low tip speed are generally considered advantageous from a hydro-acoustic point of view. Hydroacoustic noise generated directly by a propulsor can be categorized into: cavitation noise; narrowband, (or tonal), noise; and broadband noise. Each of these is discussed briefly. It is noted that if cavitation occurs it will dominate all other sources of noise. Cavitation Inception Speed is the lowest speed at which cavitation will occur. Four different operating regimes can be identified: ultra-quiet operation at low speed; normal operation at patrol speed; high speed operation; and operation close to the surface when snorkeling. Each of these is discussed briefly.

7.1 General

The propulsor is the most important hydrodynamic noise source, because the noise generation processes are speed-dependent, and the propeller or rotor blades are generally the fastest moving components in contact with the water. Thus, low rotor rotational speed (rpm) and low propulsor diameter (to reduce the rotor tip speed) are generally considered to be advantageous from a hydro-acoustic point of view.

This field is highly classified, and there is not much available in the public domain.

The hydro-acoustic noise generated directly by a propulsor can be categorised into one of the following three categories:

(a) cavitation noise;
(b) narrowband, (or tonal), noise; and
(c) broadband noise.

Cavitation is caused by the water pressure being lowered to below the vapour pressure. The collapse of the cavitation bubbles causes significant noise, and if this occurs it will dominate all other sources of noise. Cavitation Inception Speed (CIS)

M. Renilson, *Submarine Hydrodynamics*,
https://doi.org/10.1007/978-3-319-79057-2_7

is the lowest speed at which cavitation will occur. For a submarine CIS will depend on depth, being higher at deeper depths.

Generally, submarine propulsors are designed to avoid cavitation, which is possible for deeply submerged submarines due to the additional head of water above the propulsor. However, cavitation may occur when the submarine is operating close to the surface, and/or in an "off-design" condition, such as when manoeuvring.

In general, cavitation margins are defined by inception of:

(a) suction-side cavitation;
(b) suction-side tip-vortex cavitation;
(c) pressure-side cavitation;
(d) pressure-side tip-vortex cavitation; and
(e) hub-vortex cavitation.

(Anderson et al. 2009).

Strategies to ensure that each of these cavitation mechanisms is avoided are adopted in the design of a submarine propeller. These include: ensuring adequate blade area; thicker blades; and reducing the loading at the blade tip and the hub. In addition, the tip speed is reduced if possible, and the gap between the rotor tip and the duct in a pumpjet minimised.

Narrowband noise, narrowband, where the energy is focused at a number of discrete frequencies, occurs at the blade passing rate, and at harmonics of this. It is caused by the blades passing through the non-uniform wake generated by the hull and appendages ahead of the propulsor. Therefore the design of the hull and appendages ahead of the propulsor has a dominating influence on the characteristics of the narrowband noise generated by the propulsor. As noted in Chap. 5, the number of rotor blades is usually a prime number to reduce the effect of harmonics.

Broadband noise, where the energy is spread across a wide frequency range, is generated by direct turbulent interactions, and it cannot be predicted by the use of existing numerical techniques. This is because the relevant processes are represented by approximate forms, rather than modelled in their own right. Broadband noise is generated by turbulent fluctuations of the boundary layer on the blade, and is more directly associated with flow turbulence than narrowband noise.

In addition to noise generated directly by the propulsor, the fluctuating forces on the propulsor can set up resonances in the submarine, which in turn can generate radiated acoustic noise.

Low frequency noise travels further than high frequency noise. Thus, for submarine detection low frequency noise is far more critical.

At low speed very little hydrodynamic noise is produced. Machine noise may dominate, and cavitation is very unlikely. At intermediate speeds hydrodynamic flow noise will be important, both broadband and narrowband. At high speeds, particularly when running close to the surface, cavitation may occur, and when it does this dominates all noise sources.

Thus, it is clear that the relative importance of the various noise sources differs depending on the speed and operating depth of the submarine. However, detailed information about this is highly classified and not generally available in the public domain.

Four different operating regimes can be identified:

(a) ultra-quiet operation at low speed;
(b) normal operation at patrol speed;
(c) high speed operation; and
(d) operation close to the surface when snorkelling.

As the critical noise source depends on the operating regime, the mitigation processes to reduce noise for a particular submarine will depend on which regimes it operates in. For example, SSKs are generally not able to operate at speeds greater than about fifteen knots, so measures to reduce noise for submarines operating in the high speed regime are not applicable. Equally, SSNs are not required to snorkel close to the surface, so measures to reduce noise for submarines operating in this regime may not be applicable to them.

(a) Ultra-Quiet Operation at Low Speed

In the ultra-quiet mode a submarine will be travelling below about four knots. At this speed hydrodynamic noise is minimal, and machinery noise will normally be the most important noise source (Miasnikov 1995).

(b) Normal Operation at Patrol Speed

At speeds between about eight knots and fifteen knots the hydrodynamic flow noise will dominate. For well-designed deeply submerged submarines cavitation should not be an issue. Hence, the broadband and narrowband noise generated by the propulsor are the important issues for this speed range.

(c) High Speed Operation

As hydrodynamic flow noise is proportional to speed raised to the sixth power, above about fifteen knots hydrodynamic noise from the propulsor becomes very significant. Submarines operating at these speeds will be significantly more noisy than those operating at speeds below this, with the hydrodynamic noise from the propulsor being critical (Miasnikov 1995).

At these higher speeds, the fluctuating forces from the propulsor can hit hull structural modes, which may cause increased radiated acoustic noise from the submarine. An example of this is the first ‘accordion mode’, where the stern and the bow vibrate. This is far less likely to be an issue at speeds below around fifteen knots.

For a deeply submerged submarine cavitation should not be an issue until speeds of the order of twenty to twenty five knots. Above this speed, cavitation noise may dominate (Miasnikov 1995).

In addition, at speeds above about fifteen knots the hydrodynamic noise on the hull is also an issue, which may make it difficult to operate hull borne sonar.

(d) Operation Close to the Water Surface when Snorkelling

When operating close to the water surface the radiation of the hydro-acoustic noise generated can be substantial due to the presence of the water surface.

It is possible that cavitation will occur due to: the reduction in static head; the increase in drag due to the presence of the water surface and the drag on the masts resulting in increased thrust required; and the activation of the planes to control the boat close to the surface, particularly in waves.

In addition, when an SSK is snorkelling it will generate increased machinery noise.

Hydrodynamic flow noise is unlikely to be an important issue in this operating regime.

References

Andersen P, Kappel JJ, Spangenberg E (2009) Aspects of propeller developments for a submarine. In: First international symposium on marine propulsors, Trondheim, Norway

Miasnikov EV (1995) The future of Russia's strategic nuclear forces, discussions and arguments. Centre for Arms Control, Energy and Environmental Studies, Moscow Institute of Physics and Technology

Erratum to: Manoeuvring and Control

Erratum to:
Chapter 3 in: M. Renilson, *Submarine Hydrodynamics*,
https://doi.org/10.1007/978-3-319-79057-2_3

In the original version of the book, the belated corrections from author to update Eqs. (3.105)–(3.110) have to be incorporated in Chap. 3. The erratum chapter and the book have been updated with the changes.

The updated online version of this chapter can be found at
https://doi.org/10.1007/978-3-319-79057-2_3

M. Renilson, *Submarine Hydrodynamics*,
https://doi.org/10.1007/978-3-319-79057-2_8

Appendix

M. Renilson, *Submarine Hydrodynamics*,
https://doi.org/10.1007/978-3-319-79057-2

Principal features of a selection of modern submarines

Class type	Hull				Sail			Fwd planes		ACS	Propulsor	Sub-merged power (kW)	V (kn)	Comments
	L(m) D(m)	L/D	Disp (t)	Conf	Conf	l/L H/L	Loc/L	Conf	Loc/L					
Akula I SSN	110.3 13.6	8.1	12,770	Short PMB	Blended	0.23 0.03	0.10	High midline All moving	0.38	Cruciform Vertical stab with rudder Stern plane with tab	Propeller	32,000	33	Akula II & III: L = 113.3m
Alpha SSN	81.4 9.5	8.6	3200	Short PMB No casing	Blended	0.20 0.04	0.16	High midline All moving	0.44	Cruciform Vertical stab with rudder Stern plane with tab	Propeller	30,000	41	
Arihant SSBN	112.0 11.0	10.2	6000	Large casing	Foil	0.11 0.03	0.21	Sail	0.25	Cruciform Vertical stab with rudder Stern plane with tab	Propeller	83,000	24	Details scarce
Astute SSN	97.0 11.3	8.6	7600	PMB Casing	Large foil	0.17 0.06	0.13	Eyebrow All moving	0.35	Cruciform All moving rudder Stern plane with tab	Pump-jet	20,500	30	

(continued)

(continued)

Class type	Hull				Sail			Fwd planes		ACS	Propulsor	Sub-merged power (kW)	V (kn)	Comments
	L(m) D(m)	L/D	Disp (t)	Conf	Conf	l/L H/L	Loc/L	Conf	Loc/L					
Barracuda SSN	99.4 8.8	11.3	5300	PMB Very small casing	Foil	0.12 0.07	0.09	Midline All moving Retractable	0.39	X-form Fixed and movable surfaces with gap	Pump-jet	50,000	25+	
Borei SSBN	170 13.5	12.6	24,000	PMB Large casing	Foil	0.14 0.03	0.19	Midline	0.41	Cruciform Vertical stab with tab Stern plane with tab	Pump-jet	≈37,000	30	Improved Borei has all moving rudders and end plates on the stern plane
Collins SSK	77.4 7.8	9.9	3400	PMB Casing	Foil	0.17 0.08	0.12	Sail All moving	0.16	X-form All moving	Propeller	5400	20	
Columbia SSBN	171.0 13.0	13.2	20,810	PMB Casing	Foil	0.06 0.04	0.33	Sail	0.35	X-form Fixed and movable tab	Pump-jet	n/a	n/a	
Dread-nought SSBN	153.6 13.7	11.2	17,500	PMB Small casing	Foil	0.11 0.04	0.23	Midline	X	X-form	Pump-jet	n/a	n/a	
Gotland A19 SSK	60.4 6.2	9.7	1600	PMB Casing	Foil	0.14 0.08	0.18	Sail	0.20	X-form All moving	Propeller	1300	20	
Jin SSBN	135 12.5	10.8	11,000	PMB Large casing	Foil	0.11 0.05	0.29	Sail	0.32	Cruciform Vertical stab with rudder Stern plane with tab	Propeller	n/a	25	Type 094

(continued)

(continued)

Class type	Hull				Sail			Fwd planes		ACS	Propulsor	Sub-merged power (kW)	V (kn)	Comments
	L(m) D(m)	L/D	Disp (t)	Conf	Conf	l/L H/L	Loc/L	Conf	Loc/L					
Kalvari SSK	67.5 6.2	10.9	1775	PMB Small casing	Foil	0.14 0.08	0.14	Sail	0.18	Cruciform Vertical stab with rudder Stern plane with tab	Propeller	2900	20	Based on Scorpène
Kilo SSK	72.6 9.9	7.3	3950	PMB Casing	Foil	0.19 0.07	0.11	Eyebrow All moving Retractable	0.25	Cruciform but no upper rudder Rudder fixed with tab Stern plane fixed with tab	Propeller	5500	20	A number of varieties exist with slightly different lengths One fitted with PJ
Lada SSK	72.0 7.1	10.1	2700	PMB Casing	Foil	0.17 0.07	0.12	Sail All moving	0.16	Cruciform Rudder with tab Stern plane with tab	Propeller	5576	21	

(continued)

(continued)

Class type	Hull				Sail			Fwd planes		ACS	Propulsor	Sub-merged power (kW)	V (kn)	Comments
	L(m) D(m)	L/D	Disp (t)	Conf	Conf	l/L H/L	Loc/L	Conf	Loc/L					
Los Angeles SSN	110.3 10	11	6000	PMB	Foil	0.07 0.05	0.20	Sail All moving	0.22	Cruciform All moving rudder Stern plane with tab	Propeller	26,000	30+	Some had end plates on the stern planes Later, improved, class had midline planes, plus two additional fixed aft planes
Ohio SSBN/SSGN	170 13.0	13.1	18,750	PMB Casing	Foil	0.06 0.03	0.29	Sail All moving	0.29	Cruciform with end plates on stern plane All moving rudder	Propeller	45,000	25	
Rubis SSN	73.6 7.6	9.7	2600	Short PMB No casing	Foil	0.14 0.08	0.22	Sail All moving	0.25	Cruciform Vertical stab with rudder Stern plane with tab Gaps between fixed parts and movable parts	Propeller	48,000	25+	

(continued)

(continued)

Class type	Hull				Sail			Fwd planes		ACS	Propulsor	Sub-merged power (kW)	V (kn)	Comments
	L(m) D(m)	L/D	Disp (t)	Conf	Conf	l/L H/L	Loc/L	Conf	Loc/L					
Scorpène (CM) SSK	61.7 6.2	10.0	1565	PMB Small casing	Foil	0.14 0.08	0.17	Sail All moving	0.22	Cruciform Vertical stab with rudder Stern plane with tab	Propeller	2900	20	A number of varieties exist with slightly different lengths
Seawolf SSN	108 12	9.0	9200	PMB No casing	Foil	0.09 0.05	0.25	High midline. All moving	0.35	Cruciform All moving rudder Stern plane with tab Plus two fixed angled surfaces	Pump-jet	34,000	35	
Shang SSN	110.0 1.0	10.0	7000	PMB	Foil	0.12 0.06	0.23	Sail All moving	0.27	Cruciform Vertical stab with rudder Stern plane with tab	Propeller	n/a	30	Also known as Type 093A/093G Type 093 is shorter than the Type 093G Info is approx., and based on 093G

(continued)

(continued)

Class type	Hull				Sail			Fwd planes		ACS	Propulsor	Sub-merged power (kW)	V (kn)	Comments
	L(m) D(m)	L/D	Disp (t)	Conf	Conf	l/L H/L	Loc/L	Conf	Loc/L					
Shishumar SSK	67.5 6.2	10.9	1775	PMB Casing	Foil	0.15 0.07	0	Very low midline	0.40	Cruciform Upper rudder all moveable Lower rudder and stern planes fixed with tabs	Propeller	4500	22.5	Indian variation on Type 209 1500
Sind-ughosh SSK	72.6 9.9	7.3	3950	PMB Casing	Foil	0.19 0.07	0.11	Eyebrow All moving Retractable	0.25	Cruciform but no upper rudder Rudder fixed with tab Stern plane fixed with tab	Propeller	4400	17	Indian variant of Kilo class
Song SSK	74.9 8.4	8.9	2250	PMB	Foil	0.14 0.09	0.22	Sail	0.26	Cruciform Vertical stab with rudder Stern plane with tab	Propeller	n/a	22	
Soryu SSK	84.0 9.1	9.2	4200	PMB Casing	Foil	0.12 0.07	0.14	Sail All moving	0.18	X-form All moving	Propeller	6000	20	

(continued)

(continued)

Class type	Hull				Sail			Fwd planes		ACS	Propulsor	Sub-merged power (kW)	V (kn)	Comments
	L(m) D(m)	L/D	Disp (t)	Conf	Conf	l/L H/L	Loc/L	Conf	Loc/L					
Trafalgar SSN	85.4 9.8	8.7	5300	PMB Casing	Foil	0.10 0.06	0.14	Midline All moving Retractable	0.39	Cruciform All moving rudder Stern plane with tab	Pump-jet for all but first of class	11,000	30+	
Triom-phant SSBN	138 12.5	11.0	14,335	PMB Casing	Foil	0.10 0.05	0.22	Sail	0.24	Cruciform with large end plates on stern plane All moving rudder	Pump-jet	15,000	25	
Type 209 SSK	64.4 6.5	9.9	1810	PMB Large casing	Foil	0.15 0.07	0	Very low midline	0.40	Cruciform Upper rudder all moveable Lower rudder and stern plane fixed with tabs	Propeller	4500	22	Variations: 1100; 1200; 1300; 1400; 1500. Info is for 1500
Type 212 SSK	56.0 6.8	8.2	1830	PMB Large casing	Large foil/blended	0.17 0.07	0.09	Sail All moving	0.10	X-form All moving	Propeller	1700	20	

(continued)

(continued)

Class type	Hull				Sail			Fwd planes		ACS	Propulsor	Sub-merged power (kW)	V (kn)	Comments
	L(m) D(m)	L/D	Disp (t)	Conf	Conf	l/L H/L	Loc/L	Conf	Loc/L					
Type 214 SSK	65.0 6.3	10.3	1860	PMB Large upper casing and smaller lower casing	Large foil/blended	0.20 0.07	0.03	Eyebrow All moving	0.28	Cruciform All moving upper rudder. Lower rudder fixt stab plus rudder. Stern plane with tab	Propeller	2850	20	All the control surfaces are within the boundaries of the hull
Type 216 SSK	90.0 8.1	11.1	4000	PMB with casing	Large foil	0.16 0.05	0.10	Sail All moving	0.14	X-form All moving	Propeller	n/a	20+	Concept only?
Type 218SG SSK	70.0 6.3	11.1	≈2000	PMB Large upper casing and smaller lower casing	Foil	0.15 0.06	0.13	Eyebrow All moving	0.33	X-form All moving With additional fixed vertical and horizontal planes	Propeller	n/a	n/a	

(continued)

(continued)

Class type	Hull				Sail			Fwd planes		ACS	Propulsor	Sub-merged power (kW)	V (kn)	Comments
	L(m) D(m)	L/D	Disp (t)	Conf	Conf	l/L H/L	Loc/L	Conf	Loc/L					
Typhoon SSBN	175 23/12[a]	–	48,000	Non circular	Foil	0.17 0.05	−ve 0.05	Very high midline All moving	0.42	Cruciform Vertical stab with tab Stern plane with small tabs in way of propellers	Two propellers in shrouds (7)	74,000	27	[a]Non circular Sail aft of midships, with missiles forward Double hull configuration
Vanguard SSBN	149.9 12.8	11.7	15,900	PMB Casing	Foil	0.07 0.04	0.27	Eyebrow All moving	0.37	Cruciform All moving rudder Stern plane with tab	Pump-jet	20,500	25+	
Victoria SSK	70.3 7.2	9.8	2455	PMB Small casing External keel	Foil	0.16 0.08	0.08	Midline All moving	0.36	Cruciform All moving rudder Stern plane with tab	Propeller	5000	20	

(continued)

(continued)

Class type	Hull				Sail			Fwd planes		ACS	Propulsor	Sub-merged power (kW)	V (kn)	Comments
	L(m) D(m)	L/D	Disp (t)	Conf	Conf	l/L H/L	Loc/L	Conf	Loc/L					
Virginia SSN	115.0 10.0	11.5	7900	PMB No casing	Foil	0.08 0.05	0.29	High midline. All moving	0.38	Cruciform All moving rudder Stern plane with tab Plus two fixed angled surfaces Lower rudder extends below the keel line	Pump-jet	30,000	25+	
Walrus SSK	67.7 8.4	8.1	2650	Teardrop Small casing	Foil	0.18 0.09	0.17	Sail All moving At leading edge of sail	0.25	X-form All moving with headbox	Propeller	4000	20	
Xia SSBN	120 10	12	8000	PMB Large casing	Foil	0.15 0.05	0.26	Sail	0.30	Cruciform Vertical stab with tab Stern plane with tabs	Propeller	58,000	22	Type 092

(continued)

(continued)

Class type	Hull				Sail			Fwd planes		ACS	Propulsor	Sub-merged power (kW)	V (kn)	Comments
	L(m) D(m)	L/D	Disp (t)	Conf	Conf	l/L H/L	Loc/L	Conf	Loc/L					
Yasen SSN	139.5 ≈13	≈10.7	13,800	PMB	Large foil/blended	0.16 0.03	0.24	Very high midline (almost eyebrow) Retractable	0.35	Cruciform Rudder and stern plane fixed with tabs	Propeller	n/a	35	Also known as the Graney class Speed of 28 knots "silent"
Zwaardvs SSK	66.9 8.4	8.0	2640	Teardrop Small casing	Foil	0.15 0.09	0.17	Sail All moving At leading edge of sail	0.22	Cruciform All moving rudder Stern plane with tab Lower rudder extends below keel	Propeller	3800	20	

All information from public domain sources and may not be accurate
n/a indicates that the information is not available

Index

M. Renilson, *Submarine Hydrodynamics*,
https://doi.org/10.1007/978-3-319-79057-2

Printed in the United States
By Bookmasters